Fiber Reinforced Polymers:
Structural Applications

Fiber Reinforced Polymers: Structural Applications

Edited by **Jan Cooper**

New York

Published by NY Research Press,
23 West, 55th Street, Suite 816,
New York, NY 10019, USA
www.nyresearchpress.com

Fiber Reinforced Polymers: Structural Applications
Edited by Jan Cooper

International Standard Book Number: 978-1-63238-196-5 (Hardback)

Printed in the United States of America.

Contents

Preface

Every book is a source of knowledge and this one is no exception. The idea that led to the conceptualization of this book was the fact that the world is advancing rapidly; which makes it crucial to document the progress in every field. I am aware that a lot of data is already available, yet, there is a lot more to learn. Hence, I accepted the responsibility of editing this book and contributing my knowledge to the community.

The structural applications of fiber reinforced polymers are described in this elaborative book. Fiber reinforced polymers are an extremely significant and widely utilized form of material. These materials exist primarily due to the emphasis given to petrochemical and non-petrochemical products. These polymers are economical and more widely used as compared to natural polymers. The field of fiber reinforced polymers is an evolving and growing field of research due to its special properties and traditional and modern applications. This book discusses the improvement in environmental aspects, important areas of research and potentials of fiber reinforced polymers in civil construction and concrete repair.

While editing this book, I had multiple visions for it. Then I finally narrowed down to make every chapter a sole standing text explaining a particular topic, so that they can be used independently. However, the umbrella subject sinews them into a common theme. This makes the book a unique platform of knowledge.

I would like to give the major credit of this book to the experts from every corner of the world, who took the time to share their expertise with us. Also, I owe the completion of this book to the never-ending support of my family, who supported me throughout the project.

Editor

Basics Concepts of Polymers Used in FRP

Introduction of Fibre-Reinforced Polymers – Polymers and Composites: Concepts, Properties and Processes

Martin Alberto Masuelli

Additional information is available at the end of the chapter

1. Introduction

Fibre-reinforced polymer(FRP), also *Fibre-reinforced plastic*, is a composite material made of a polymer matrix reinforced with fibres. The fibres are usually glass, carbon, or aramid, although other fibres such as paper or wood or asbestos have been sometimes used. The polymer is usually an epoxy, vinylester or polyester thermosetting plastic, and phenol formaldehyde resins are still in use. FRPs are commonly used in the aerospace, automotive, marine, and construction industries.

Composite materials are engineered or naturally occurring materials made from two or more constituent materials with significantly different physical or chemical properties which remain separate and distinct within the finished structure. Most composites have strong, stiff fibres in a matrix which is weaker and less stiff. The objective is usually to make a component which is strong and stiff, often with a low density. Commercial material commonly has glass or carbon fibres in matrices based on thermosetting polymers, such as epoxy or polyester resins. Sometimes, thermoplastic polymers may be preferred, since they are moldable after initial production. There are further classes of composite in which the matrix is a metal or a ceramic. For the most part, these are still in a developmental stage, with problems of high manufacturing costs yet to be overcome [1]. Furthermore, in these composites the reasons for adding the fibres (or, in some cases, particles) are often rather complex; for example, improvements may be sought in creep, wear, fracture toughness, thermal stability, etc [2].

Fibre reinforced polymer (FRP) are composites used in almost every type of advanced engineering structure, with their usage ranging from aircraft, helicopters and spacecraft through to boats, ships and offshore platforms and to automobiles, sports goods, chemical processing equipment and civil infrastructure such as bridges and buildings. The usage of FRP composites continues to grow at an impressive rate as these materials are used more in their existing markets and become established in relatively new markets such as biomedical devices and civil structures. A key factor driving the increased applications of composites over the recent years is the development of new advanced forms of FRP materials. This includes developments in high performance resin systems and new styles of reinforcement, such as carbon nanotubes and nanoparticles. This book provides an up-to-date account of the fabrication, mechanical properties, delamination resistance, impact tolerance and applications of 3D FRP composites [3].

The fibre reinforced polymer composites (FRPs) are increasingly being considered as an enhancement to and/or substitute for infrastructure components or systems that are constructed of traditional civil engineering materials, namely concrete and steel. FRP composites are lightweight, no-corrosive, exhibit high specific strength and specific stiffness, are easily constructed, and can be tailored to satisfy performance requirements. Due to these advantageous characteristics, FRP composites have been included in new construction and rehabilitation of structures through its use as reinforcement in concrete, bridge decks, modular structures, formwork, and external reinforcement for strengthening and seismic upgrade [4].

The applicability of Fiber Reinforced Polymer (FRP) reinforcements to concrete structures as a substitute for steel bars or prestressing tendons has been actively studied in numerous research laboratories and professional organizations around the world. FRP reinforcements offer a number of advantages such as corrosion resistance, non-magnetic properties, high tensile strength, lightweight and ease of handling. However, they generally have a linear elastic response in tension up to failure (described as a brittle failure) and a relatively poor transverse or shear resistance. They also have poor resistance to fire and when exposed to high temperatures. They loose significant strength upon bending, and they are sensitive to stress-rupture effects. Moreover, their cost, whether considered per unit weight or on the basis of force carrying capacity, is high in comparison to conventional steel reinforcing bars or prestressing tendons. From a structural engineering viewpoint, the most serious problems with FRP reinforcements are the lack of plastic behavior and the very low shear strength in the transverse direction. Such characteristics may lead to premature tendon rupture, particularly when combined effects are present, such as at shear-cracking planes in reinforced concrete beams where dowel action exists. The dowel action reduces residual tensile and shear resistance in the tendon. Solutions and limitations of use have been offered and continuous improvements are expected in the future. The unit cost of FRP reinforcements is expected to decrease significantly with increased market share and demand. However, even today, there are applications where FRP reinforcements are cost effective and justifiable. Such cases include the use of bonded FRP sheets or plates in repair and strengthening of concrete structures, and the use of FRP meshes or textiles or fabrics in thin cement products. The cost of repair and rehabilitation of a structure is always, in relative terms, substantially higher than the cost of the initial structure. Repair generally requires a relatively small volume of repair materials but a relatively high commitment in labor. Moreover the cost of labor in developed countries is so high that the cost of material becomes secondary. Thus the highest the performance and durability of the repair material is, the more cost-effective is the repair. This implies that material cost is not really an issue in repair and that the fact that FRP repair materials are costly is not a constraining drawback [5].

When considering only energy and material resources it appears, on the surface, the argument for FRP composites in a sustainable built environment is questionable. However, such a conclusion needs to be evaluated in terms of potential advantages present in use of FRP composites related to considerations such as:

• Higher strength

• Lighter weight

• Higher performance

• Longer lasting

• Rehabilitating existing structures and extending their life

• Seismic upgrades

• Defense systems

• Space systems

• Ocean environments

In the case of FRP composites, environmental concerns appear to be a barrier to its feasibility as a sustainable material especially when considering fossil fuel depletion, air pollution, smog, and acidification associated with its production. In addition, the ability to recycle FRP composites is limited and, unlike steel and timber, structural components cannot be reused to perform a similar function in another structure. However, evaluating the environmental impact of FRP composites in infrastructure applications, specifically through life cycle analysis, may reveal direct and indirect benefits that are more competitive than conventional materials.

Composite materials have developed greatly since they were first introduced. However, before composite materials can be used as an alternative to conventional materials as part of a sustainable environment a number of needs remain.

• Availability of standardized durability characterization data for FRP composite materials.

• Integration of durability data and methods for service life prediction of structural members utilizing FRP composites.

• Development of methods and techniques for materials selection based on life cycle assessments of structural components and systems.

Ultimately, in order for composites to truly be considered a viable alternative, they must be structurally and economically feasible. Numerous studies regarding the structural feasibility of composite materials are widely available in literature [6]. However, limited studies are available on the economic and environmental feasibility of these materials from the perspective of a life cycle approach, since short term data is available or only economic costs are considered in the comparison. Additionally, the long term affects of using composite materials needs to be determined. The byproducts of the production, the sustainability of the constituent materials, and the potential to recycle composite materials needs to be assessed in order to determine of composite materials can be part of a sustainable environment. Therefore in this chapter describe the physicochemical properties of polymers and composites more used in Civil Engineering. The theme will be addressed in a simple and basic for better understanding.

2. Manufactured process and basic concepts

The synthetic polymers are generally manufactured by polycondensation, polymerization or polyaddition. The polymers combined with various agents to enhance or in any way alter the material properties of polymers the result is referred to as a plastic. The Composite plastics can be of homogeneous or heterogeneous mix. Composite plastics refer to those types of plastics that result from bonding two or more homogeneous materials with different material properties to derive a final product with certain desired material and mechanical properties. The Fibre reinforced plastics (or fiber reinforced polymers) are a category of composite plastics that specifically use fibre materials (not mix with polymer) to mechanically enhance the strength and elasticity of plastics. The original plastic material without fibre reinforcement is known as the matrix. The matrix is a tough but relatively weak plastic that is reinforced by stronger stiffer reinforcing filaments or fibres. The extent that strength and elasticity are enhanced in a fibre reinforced plastic depends on the mechanical properties of the fibre and matrix, their volume relative to one another, and the fibre length and orientation within the matrix. Reinforcement of the matrix occurs by definition when the FRP material exhibits increased strength or elasticity relative to the strength and elasticity of the matrix alone.

Polymers are different from other construction materials like ceramics and metals, because of their macromolecular nature. The covalently bonded, long chain structure makes them macromolecules and determines, via the weight averaged molecular weight, Mw, their processability, like spin-, blow-, deep draw-, generally melt-formability. The number averaged molecular weight, Mn, determines the mechanical strength, and high molecular weights are beneficial for properties like strain-to-break, impact resistance, wear, etc. Thus, natural limits are met, since too high molecular weights yield too high shear and elongational viscosities that make polymers inprocessable. Prime examples are the very useful poly-tetra-fluor-ethylenes, PTFE's, and ultrahigh-molecular-weight-poly-ethylenes, UHMWPE's, and not only garbage bags are made of polyethylene, PE, but also high-performance fibers that are even used for bullet proof vests (alternatively made from, also inprocessable in the melt, rigid aromatic polyamides). The resulting mechanical properties of these high performance fibers, with moduli of 150 GPa and strengths of up to 4 GPa, represent the optimal use of what the potential of the molecular structure of polymers yields, combined with their low density. Thinking about polymers, it becomes clear why living nature used the polymeric concept to build its structures, and not only in high strength applications like wood, silk or spider-webs [7].

2.1. Polymers

The linking of small molecules (monomers) to make larger molecules is a polymer. Polymerization requires that each small molecule have at least two reaction points or functional groups. There are two distinct major types of polymerization processes, condensation polymerization, in which the chain growth is accompanied by elimination of small molecules such as H_2O or CH_3OH, and addition polymerization, in which the polymer is formed without the loss of other materials. There are many variants and subclasses of polymerization reactions.

The polymer chains can be classified in linear polymer chain, branched polymer chain, and cross-linked polymer chain. The structure of the repeating unit is the difunctional monomeric unit, or "mer." In the presence of catalysts or initiators, the monomer yields a polymer by the joining together of n-mers. If n is a small number, 2–10, the products are dimers, trimers, tetramers, or oligomers, and the materials are usually gases, liquids, oils, or brittle solids. In most solid polymers, n has values ranging from a few score to several hundred thousand, and the corresponding molecular weights range from a few thousand to several million. The end groups of this example of addition polymers are shown to be fragments of the initiator. If only one monomer is polymerized, the product is called a homopolymer. The polymerization of a mixture of two monomers of suitable reactivity leads to the formation of a copolymer, a polymer in which the two types of mer units have entered the chain in a more or less random fashion. If chains of one homopolymer are chemically joined to chains of another, the product is called a block or graft copolymer.

Isotactic and syndiotactic (stereoregular) polymers are formed in the presence of complex catalysts, or by changing polymerization conditions, for example, by lowering the temperature. The groups attached to the chain in a stereoregular polymer are in a spatially ordered arrangement. The regular structures of the isotactic and syndiotactic forms make them often capable of crystallization. The crystalline melting points of isotactic polymers are often substantially higher than the softening points of the atactic product.

The spatially oriented polymers can be classified in atactic (random; dlldl or lddld, and so on), syndiotactic (alternating; dldl, and so on), and isotactic (right- or left-handed; dddd, or llll, and so on). For illustration, the heavily marked bonds are assumed to project up from the paper, and the dotted bonds down. Thus in a fully syndiotactic polymer, asymmetric carbons alternate in their left- or right-handedness (alternating d, l configurations), while in an isotactic polymer, successive carbons have the same steric configuration (d or l). Among the several kinds of polymerization catalysis, free-radical initiation has been most thoroughly studied and is most widely employed. Atactic polymers are readily formed by free-radical polymerization, at moderate temperatures, of vinyl and diene monomers and some of their derivatives. Some polymerizations can be initiated by materials, often called ionic catalysts, which contain highly polar reactive sites or complexes. The term heterogeneous catalyst is often applicable to these materials because many of the catalyst systems are insoluble in monomers and other solvents. These polymerizations are usually carried out in solution from which the polymer can be obtained by evaporation of the solvent or by precipitation on the addition of a nonsolvent. A distinguishing feature of complex catalysts is the ability of some representatives of each type to initiate stereoregular polymerization at ordinary temperatures or to cause the formation of polymers which can be crystallized [1, 6].

2.1.1. Polymerization

Polymerization, emulsion polymerization any process in which relatively small molecules, called monomers, combine chemically to produce a very large chainlike or network molecule, called a polymer. The monomer molecules may be all alike, or they may represent two, three, or more different compounds. Usually at least 100 monomer molecules must be combined to make a product that has certain unique physical properties-such as elasticity, high tensile strength, or the ability to form fibres-that differentiate polymers from substances composed of smaller and simpler molecules; often, many thousands of monomer units are incorporated in a single molecule of a polymer. The formation of stable covalent chemical bonds between the monomers sets polymerization apart from other processes, such as crystallization, in which large numbers of molecules aggregate under the influence of weak intermolecular forces.

Two classes of polymerization usually are distinguished. In condensation polymerization, each step of the process is accompanied by formation of a molecule of some simple compound, often water. In addition polymerization, monomers react to form a polymer without the formation of by-products. Addition polymerizations usually are carried out in the presence of catalysts, which in certain cases exert control over structural details that have important effects on the properties of the polymer [8].

Linear polymers, which are composed of chainlike molecules, may be viscous liquids or solids with varying degrees of crystallinity; a number of them can be dissolved in certain liquids, and they soften or melt upon heating. Cross-linked polymers, in which the molecular structure is a network, are thermosetting resins (i.e., they form under the influence of heat but, once formed, do not melt or soften upon reheating) that do not dissolve in solvents. Both linear and cross-linked polymers can be made by either addition or condensation polymerization.

2.1.2. Polycondensation

The polycondensation a process for the production of polymers from bifunctional and polyfunctional compounds (monomers), accompanied by the elimination of low-molecular weight by-products (for example, water, alcohols, and hydrogen halides). A typical example of polycondensation is the synthesis of complex polyester.

The process is called homopolycondensation if the minimum possible number of monomer types for a given case participates, and this number is usually two. If at least one monomer more than the number required for the given reaction participates in polycondensation, the process is called copolycondensation. Polycondensation in which only bifunctional com pounds participate leads to the formation of linear macromolecules and is called linear poly-condensation. If molecules with three or more functional groups participate in polycondensation, three-dimensional structures are formed and the process is called three-dimensional polycondensation. In cases where the degree of completion of polycondensation and the mean length of the macromolecules are limited by the equilibrium concentration of the reagents and reaction products, the process is called equilibrium (reversible) polycondensation. If the limiting factors are kinetic rather than thermodynamic, the process is called nonequilibrium (irreversible) polycondensation.

Polycondensation is often complicated by side reactions, in which both the original monomers and the polycondensation products (oligomers and polymers) may participate. Such reactions include the reaction of monomer or oligomer with a mono-functional compound (which may be present as an impurity), intramolecular cyclization (ring closure), and degradation of the macromolecules of the resultant polymer. The rate competition of polycondensation and the side reactions determines the molecular weight, yield, and molecular weight distribution of the polycondensation polymer.

Polycondensation is characterized by disappearance of the monomer in the early stages of the process and a sharp increase in molecular weight, in spite of a slight change in the extent of conversion in the region of greater than 95-percent conversion.

A necessary condition for the formation of macro-molecular polymers in linear polycondensation is the equivalence of the initial functional groups that react with one another.

Polycondensation is accomplished by one of three methods:

1. in a melt, when a mixture of the initial compounds is heated for a long period to 10°-20°C above the melting (softening) point of the resultant polymer;

2. in solution, when the monomers are present in the same phase in the solute state;

3. on the phase boundary between two immiscible liquids, in which one of the initial compounds is found in each of the liquid phases (interphase polycondensation).

Polycondensation processes play an important role in nature and technology. Polycondensation or similar reactions are the basis for the biosynthesis of the most important biopolymers-proteins, nucleic acids, and cellulose. Polycondensation is widely used in industry for the production of polyesters (polyethylene terephthalate, polycarbonates, and alkyd resins), polyamides, phenol-formaldehyde resins, urea-formaldehyde resins, and certain silicones [9]. In the period 1965-70, polycondensation acquired great importance in connection with the development of industrial production of a series of new polymers, including heat-resistant polymers (polyarylates, aromatic polyimides, polyphe-nylene oxides, and polysulfones).

2.1.3. Polyaddition

The polyaddition reactions are similar to polycondensation reactions because they are also step reactions, however without splitting off low molecular weight by-products. The reaction is exothermic rather than endothermic and therefore cannot be stopped at will. Typical for polyaddition reaction is that individual atoms, usually H-atoms, wander from one monomer to another as the two monomers combine through a covalent bond. The monomers, as in polycondensation reactions, have to be added in stoichiometric amounts. These reactions do not start spontaneously and they are slow.

Polyaddition does not play a significant role in the production of thermoplastics. It is commonly encountered with cross-linked polymers. Polyurethane, which can be either a thermoplastic or thermosets, is synthesized by the reaction of multi-functional isocyanates with multifunctional amines or alcohol. Thermosetting epoxy resins are formed by polyaddition of epoxides with curing agents, such as amines and acid anhydrides.

In comparing chain reaction polymerization with the other two types of polymerization the following principal differences should be noted: Chain reaction polymerization, or simply called polymerization, is a chain reaction as the name implies. Only individual monomer molecules add to a reactive growing chain end, except for recombination of two radical chain ends or reactions of a reactive chain end with an added modifier molecule. The activation energy for chain initiation is much grater than for the subsequent growth reaction and growth, therefore, occurs very rapidly.

2.2. Composites

Composite is any material made of more than one component. There are a lot of composites around you. Concrete is a composite. It's made of cement, gravel, and sand, and often has steel rods inside to reinforce it. Those shiny balloons you get in the hospital when you're sick are made of a composite, which consists of a polyester sheet and an aluminum foil sheet, made into a sandwich. The polymer composites made from polymers, or from polymers along with other kinds of materials [7]. But specifically the fiber-reinforced composites are materials in which a fiber made of one material is embedded in another material.

2.2.1. Polymer composites

The polymer composites are any of the combinations or compositions that comprise two or more materials as separate phases, at least one of which is a polymer. By combining a polymer with another material, such as glass, carbon, or another polymer, it is often possible to obtain unique combinations or levels of properties. Typical examples of synthetic polymeric composites include glass-, carbon-, or polymer-fiber-reinforced, thermoplastic or thermosetting resins, carbon-reinforced rubber, polymer blends, silica- or mica-reinforced resins, and polymer-bonded or -impregnated concrete or wood. It is also often useful to consider as composites such materials as coatings (pigment-binder combinations) and crystalline polymers (crystallites in a polymer matrix). Typical naturally occurring composites include wood (cellulosic fibers bonded with lignin) and bone (minerals bonded with collagen). On the other hand, polymeric compositions compounded with a plasticizer or very low proportions of pigments or processing aids are not ordinarily considered as composites.

Typically, the goal is to improve strength, stiffness, or toughness, or dimensional stability by embedding particles or fibers in a matrix or binding phase. A second goal is to use inexpen sive, readily available fillers to extend a more expensive or scarce resin; this goal is increasingly important as petroleum supplies become costlier and less reliable. Still other applications include the use of some filler such as glass spheres to improve processability, the incorporation of dry-lubricant particles such as molybdenum sulfide to make a self-lubricating bearing, and the use of fillers to reduce permeability.

The most common fiber-reinforced polymer composites are based on glass fibers, cloth, mat, or roving embedded in a matrix of an epoxy or polyester resin. Reinforced thermosetting resins containing boron, polyaramids, and especially carbon fibers confer especially high levels of strength and stiffness. Carbon-fiber composites have a relative stiffness five times that of steel. Because of these excellent properties, many applications are uniquely suited for epoxy and polyester composites, such as components in new jet aircraft, parts for automobiles, boat hulls, rocket motor cases, and chemical reaction vessels.

Although the most dramatic properties are found with reinforced thermosetting resins such as epoxy and polyester resins, significant improvements can be obtained with many reinforced thermoplastic resins as well. Polycarbonates, polyethylene, and polyesters are among the resins available as glass-reinforced composition. The combination of inexpensive, one-step fabrication by injection molding, with improved properties has made it possible for reinforced thermoplastics to replace metals in many applications in appliances, instruments, automobiles, and tools.

In the development of other composite systems, various matrices are possible; for example, polyimide resins are excellent matrices for glass fibers, and give a high- performance composite. Different fibers are of potential interest, including polymers [such as poly(vinyl alcohol)], single-crystal ceramic whiskers (such as sapphire), and various metallic fibers.

Long ago, people living in South and Central America had used natural rubber latex, polyisoprene, to make things like gloves and boots, as well as rubber balls which they used to play games that were a lot like modern basketball. He took two layers of cotton fabric and embedded them in natural rubber, also known as polyisoprene, making a three-layered sandwich like the one you see on your right (Remember, cotton is made up of a natural polymer called cellulose). This made for good raincoats because, while the rubber made it waterproof, the cotton layers made it comfortable to wear, to make a material that has the properties of both its components. In this case, we combine the water-resistance of polyisoprene and the comfort of cotton.

Modern composites are usually made of two components, a fiber and matrix. The fiber is most often glass, but sometimes Kevlar, carbon fiber, or polyethylene. The matrix is usually a thermoset like an epoxy resin, polydicyclopentadiene, or a polyimide. The fiber is embedded in the matrix in order to make the matrix stronger. Fiber-reinforced composites have two things going for them. They are strong and light. They are often stronger than steel, but weigh much less. This means that composites can be used to make automobiles lighter, and thus much more fuel efficient.

A common fiber-reinforced composite is Fiberglas™. Its matrix is made by reacting polyester with carbon-carbon double bonds in its backbone, and styrene. We pour a mix of the styrene and polyester over a mass of glass fibers.

The styrene and the double bonds in the polyester react by free radical vinyl polymerization to form a crosslinked resin. The glass fibers are trapped inside, where they act as a reinforcement. In Fiberglas™ the fibers are not lined up in any particular direction. They are just a tangled mass, like you see on the right. But we can make the composite stronger by lining up all the fibers in the same direction. Oriented fibers do some weird things to the composite. When you pull on the composite in the direction of the fibers, the composite is very strong. But if you pull on it at right angles to the fiber direction, it is not very strong at all [8-9]. This is not always bad, because sometimes we only need the composite to be strong in one direction. Sometimes the item you are making will only be under stress in one direction. But sometimes we need strength in more than one direction. So we simply point the fibers in more than one direction. We often do this by using a woven fabric of the fibers to reinforce the composite. The woven fibers give a composite good strength in many directions.

The polymeric matrix holds the fibers together. A loose bundle of fibers would not be of much use. Also, though fibers are strong, they can be brittle. The matrix can absorb energy by deforming under stress. This is to say, the matrix adds toughness to the composite. And finally, while fibers have good tensile strength (that is, they are strong when you pull on them), they usually have awful compressional strength. That is, they buckle when you squash them. The matrix gives compressional strength to the composite.

Not all fibers are the same. Now it may seem strange that glass is used as reinforcement, as glass is really easy to break. But for some reason, when glass is spun into really tiny fibers, it acts very different. Glass fibers are strong, and flexible.

Still, there are stronger fibers out there. This is a good thing, because sometimes glass just isn't strong and tough enough. For some things, like airplane parts, that undergo a lot of stress, you need to break out the fancy fibers. When cost is no object, you can use stronger, but more expensive fibers, like Kevlar™, carbon fiber. Carbon fiber (Spectra™) is usually stronger than Kevlar™, that is, it can withstand more force without breaking. But Kevlar™ tends to be tougher. This means it can absorb more energy without breaking. It can stretch a little to keep from breaking, more so than carbon fiber can. But Spectra™, which is a kind of polyethylene, is stronger and tougher than both carbon fiber and Kevlar™.

Different jobs call for different matrices. The unsaturated polyester/styrene systems at are one example. They are fine for everyday applications. Chevrolet Corvette bodies are made from composites using unsaturated polyester matrices and glass fibers. But they have some drawbacks. They shrink a good deal when they're cured, they can absorb water very easily, and their impact strength is low.

2.2.2. Biocomposites

For many decades, the residential construction field has used timber as its main source of building material for the frames of modern American homes. The American timber industry produced a record 49.5 billion board feet of lumber in 1999, and another 48.0 billion board feet in 2002. At the same time that lumber production is peaking, the home ownership rate reached a record high of 69.2%, with over 977,000 homes being sold in 2002. Because residential construction accounts for one-third of the total softwood lumber use in the United States, there is an increasing demand for alternate materials. Use of sawdust not only provides an alternative but also increases the use of the by product efficiently. Wood plastic composites (WPC) is a relatively new category of materials that covers a broad range of composite materials utilizing an organic resin binder (matrix) and fillers composed of cellulose materials. The new and rapidly developing biocomposite materials are high technology products, which have one unique advantage – the wood filler can include sawdust and scrap wood products. Consequently, no additional wood resources are needed to manufacture biocomposites. Waste products that would traditraditionally cost money for proper disposal, now become a beneficial resource, allowing recycling to be both profitable and environmentally conscious. The use of biocomposites and WPC has increased rapidly all over the world, with the end users for these composites in the construction, motor vehicle, and furniture industries. One of the primary problems related to the use of biocomposites is the flammability of the two main components (binder and filler). If a flame retardant were added, this would require the adhesion of the fiber and the matrix not to be disturbed by the retardant. The challenge is to develop a composite that will not burn and will maintain its level of mechanical performance. In lieu of organic matrix compounds, inorganic matrices can be utilized to improve the fire resistance. Inorganic-based wood composites are those that consist of a mineral mix as the binder system. Such inorganic binder systems include gypsum and Portland cement, both of which are highly resistant to fire and insects. The main disadvantage with these systems is the maximum amount of sawdust or fibers than can be incorporated is low. One relatively new type of inorganic matrix is potassium aluminosilicate, an environmentally friendly compound made from naturally occurring materials. The Federal Aviation Administration has investigated the feasibility of using this matrix in commercial aircraft due to its ability to resist temperatures of up to 1000 ºC without generating smoke, and its ability to enable carbon composites to withstand temperatures of 800 ºC and maintain 63% of its original flexural strength. Potassium aluminosilicate matrices are compatible with many common building material including clay brick, masonry, concrete, steel, titanium, balsa, oak, pine, and particleboard [10].

2.3. Fiberglass

Fiberglass refers to a group of products made from individual glass fibers combined into a variety of forms. Glass fibers can be divided into two major groups according to their geometry: continuous fibers used in yarns and textiles, and the discontinuous (short) fibers used as batts, blankets, or boards for insulation and filtration. Fiberglass can be formed into yarn much like wool or cotton, and woven into fabric which is sometimes used for draperies. Fiberglass textiles are commonly used as a reinforcement material for molded and laminated plastics. Fiberglass wool, a thick, fluffy material made from discontinuous fibers, is used for thermal insulation and sound absorption. It is commonly found in ship and submarine bulkheads and hulls; automobile engine compartments and body panel liners; in furnaces and air conditioning units; acoustical wall and ceiling panels; and architectural partitions. Fiberglass can be tailored for specific applications such as Type E (electrical), used as electrical insulation tape, textiles and reinforcement; Type C (chemical), which has superior acid resistance, and Type T, for thermal insulation [11].

Though commercial use of glass fiber is relatively recent, artisans created glass strands for decorating goblets and vases during the Renaissance. A French physicist, Rene-Antoine Ferchault de Reaumur, produced textiles decorated with fine glass strands in 1713. Glass wool, a fluffy mass of discontinuous fiber in random lengths, was first produced in Europe in 1900, using a process that involved drawing fibers from rods horizontally to a revolving drum [12].

The basic raw materials for fiberglass products are a variety of natural minerals and manufactured chemicals. The major ingredients are silica sand, limestone, and soda ash. Other ingredients may include calcined alumina, borax, feldspar, nepheline syenite, magnesite, and kaolin clay, among others. Silica sand is used as the glass former, and soda ash and limestone help primarily to lower the melting temperature. Other ingredients are used to improve certain properties, such as borax for chemical resistance. Waste glass, also called cullet, is also used as a raw material. The raw materials must be carefully weighed in exact quantities and thoroughly mixed together (called batching) before being melted into glass.

2.3.1. The manufacturing process

2.3.1.1. Melting

Once the batch is prepared, it is fed into a furnace for melting. The furnace may be heated by electricity, fossil fuel, or a combination of the two. Temperature must be precisely controlled to maintain a smooth, steady flow of glass. The molten glass must be kept at a higher temperature (about 1371 °C) than other types of glass in order to be formed into fiber. Once the glass becomes molten, it is transferred to the forming equipment via a channel (forehearth) located at the end of the furnace [13].

2.3.1.2. Forming into fibers

Several different processes are used to form fibers, depending on the type of fiber. Textile fibers may be formed from molten glass directly from the furnace, or the molten glass may be fed first to a machine that forms glass marbles of about 0.62 inch (1.6 cm) in diameter. These marbles allow the glass to be inspected visually for impurities. In both the direct melt and marble melt process, the glass or glass marbles are fed through electrically heated bushings (also called spinnerets). The bushing is made of platinum or metal alloy, with anywhere from 200 to 3,000 very fine orifices. The molten glass passes through the orifices and comes out as fine filaments [13].

2.3.1.3. Continuous-filament process

A long, continuous fiber can be produced through the continuous-filament process. After the glass flows through the holes in the bushing, multiple strands are caught up on a high-speed winder. The winder revolves at about 3 km a minute, much faster than the rate of flow from the bushings. The tension pulls out the filaments while still molten, forming strands a fraction of the diameter of the openings in the bushing. A chemical binder is applied, which helps keep the fiber from breaking during later processing. The filament is then wound onto tubes. It can now be twisted and plied into yarn [14].

2.3.1.4. Staple-fiber process

An alternative method is the staplefiber process. As the molten glass flows through the bushings, jets of air rapidly cool the filaments. The turbulent bursts of air also break the filaments into lengths of 20-38 cm. These filaments fall through a spray of lubricant onto a revolving drum, where they form a thin web. The web is drawn from the drum and pulled into a continuous strand of loosely assembled fibers [15]. This strand can be processed into yarn by the same processes used for wool and cotton.

2.3.1.5. Chopped fiber

Instead of being formed into yarn, the continuous or long-staple strand may be chopped into short lengths. The strand is mounted on a set of bobbins, called a creel, and pulled through a machine which chops it into short pieces. The chopped fiber is formed into mats to which a binder is added. After curing in an oven, the mat is rolled up. Various weights and thicknesses give products for shingles, built-up roofing, or decorative mats [16].

2.3.1.6. Glass wool

The rotary or spinner process is used to make glass wool. In this process, molten glass from the furnace flows into a cylindrical container having small holes. As the container spins rapidly, horizontal streams of glass flow out of the holes. The molten glass streams are converted into fibers by a downward blast of air, hot gas, or both. The fibers fall onto a conveyor belt, where they interlace with each other in a fleecy mass. This can be used for insulation, or the wool can be sprayed with a binder, compressed into the desired thickness, and cured in an oven. The heat sets the binder, and the resulting product may be a rigid or semi-rigid board, or a flexible bat [15-16].

2.3.1.7. Protective coatings

In addition to binders, other coatings are required for fiberglass products. Lubricants are used to reduce fiber abrasion and are either directly sprayed on the fiber or added into the binder. An anti-static composition is also sometimes sprayed onto the surface of fiberglass insulation mats during the cooling step. Cooling air drawn through the mat causes the anti-static agent to penetrate the entire thickness of the mat. The anti-static agent consists of two ingredients a material that minimizes the generation of static electricity, and a material that serves as a corrosion inhibitor and stabilizer.

Sizing is any coating applied to textile fibers in the forming operation, and may contain one or more components (lubricants, binders, or coupling agents). Coupling agents are used on strands that will be used for reinforcing plastics, to strengthen the bond to the reinforced material. Sometimes a finishing operation is required to remove these coatings, or to add another coating. For plastic reinforcements, sizings may be removed with heat or chemicals and a coupling agent applied. For decorative applications, fabrics must be heat treated to remove sizings and to set the weave. Dye base coatings are then applied before dying or printing [15-16].

2.3.1.8. Forming into shapes

Fiberglass products come in a wide variety of shapes, made using several processes. For example, fiberglass pipe insulation is wound onto rod-like forms called mandrels directly from the forming units, prior to curing. The mold forms, in lengths of 91 cm or less, are then cured in an oven. The cured lengths are then de-molded lengthwise, and sawn into specified dimensions. Facings are applied if required, and the product is packaged for shipment [17].

2.4. Carbon fibre

Carbon-fiber-reinforced polymer or carbon-fiber-reinforced plastic (CFRP or CRP or often simply carbon fiber), is a very strong and light fiber-reinforced polymer which contains carbon fibers. Carbon fibres are created when polyacrylonitrile fibres (PAN), Pitch resins, or Rayon are carbonized (through oxidation and thermal pyrolysis) at high temperatures. Through further processes of graphitizing or stretching the fibres strength or elasticity can be enhanced respectively. Carbon fibres are manufactured in diameters analogous to glass fibres with diameters ranging from 9 to 17 μm. These fibres wound into larger threads for transportation and further production processes. Further production processes include weaving or braiding into carbon fabrics, cloths and mats analogous to those described for glass that can then be used in actual reinforcement processes. Carbon fibers are a new breed of high-strength materials. Carbon fiber has been described as a fiber containing at least 90% carbon obtained by the controlled pyrolysis of appropriate fibers. The existence of carbon fiber came into being in 1879 when Edison took out a patent for the manufacture of carbon filaments suitable for use in electric lamps [18].

2.4.1. Classification and types

Based on modulus, strength, and final heat treatment temperature, carbon fibers can be classified into the following categories:

1.   Based on carbon fiber properties, carbon fibers can be grouped into:

• Ultra-high-modulus, type UHM (modulus >450Gpa)

- High-modulus, type HM (modulus between 350-450Gpa)

- Intermediate-modulus, type IM (modulus between 200-350Gpa)

- Low modulus and high-tensile, type HT (modulus < 100Gpa, tensile strength > 3.0Gpa)

- Super high-tensile, type SHT (tensile strength > 4.5Gpa)

2.   Based on precursor fiber materials, carbon fibers are classified into;

- PAN-based carbon fibers

- Pitch-based carbon fibers

- Mesophase pitch-based carbon fibers

- Isotropic pitch-based carbon fibers

- Rayon-based carbon fibers

- Gas-phase-grown carbon fibers

3.   Based on final heat treatment temperature, carbon fibers are classified into:

- Type-I, high-heat-treatment carbon fibers (HTT), where final heat treatment temperature should be above 2000°C and can be associated with high-modulus type fiber.

- Type-II, intermediate-heat-treatment carbon fibers (IHT), where final heat treatment temperature should be around or above 1500 °C and can be associated with high-strength type fiber.

- Type-III, low-heat-treatment carbon fibers, where final heat treatment temperatures not greater than 1000 °C. These are low modulus and low strength materials [19].

2.4.2. Manufacture

In Textile Terms and Definitions, carbon fiber has been described as a fiber containing at least 90% carbon obtained by the controlled pyrolysis of appropriate fibers. The term "graphite fiber" is used to describe fibers that have carbon in excess of 99%. Large varieties of fibers called precursors are used to produce carbon fibers of different morphologies and different specific characteristics. The most prevalent precursors are polyacrylonitrile (PAN), cellulosic fibers (viscose rayon, cotton), petroleum or coal tar pitch and certain phenolic fibers.

Carbon fibers are manufactured by the controlled pyrolysis of organic precursors in fibrous form. It is a heat treatment of the precursor that removes the oxygen, nitrogen and hydrogen to form carbon fibers. It is well established in carbon fiber literature that the mechanical properties of the carbon fibers are improved by increasing the crystallinity and orientation, and by reducing defects in the fiber. The best way to achieve this is to start with a highly oriented precursor and then maintain the initial high orientation during the process of stabilization and carbonization through tension [18-19].

2.4.2.1. Carbon fibers from polyacrylonitrile (PAN)

There are three successive stages in the conversion of PAN precursor into high-performance carbon fibers. Oxidative stabilization: The polyacrylonitrile precursor is first stretched and simultaneously oxidized in a temperature range of 200-300 °C. This treatment converts thermoplastic PAN to a non-plastic cyclic or ladder compound. Carbonization: After oxidation, the fibers are carbonized at about 1000 °C without tension in an inert atmosphere (normally nitrogen) for a few hours. During this process the non-carbon elements are removed as volatiles to give carbon fibers with a yield of about 50% of the mass of the original PAN. Graphitization: Depending on the type of fiber required, the fibers are treated at temperatures between 1500-3000 °C, which improves the ordering, and orientation of the crystallites in the direction of the fiber axis.

2.4.2.2. Carbon fibers from rayon

a- The conversion of rayon fibers into carbon fibers is three phase process

Stabilization: Stabilization is an oxidative process that occurs through steps. In the first step, between 25-150 °C, there is physical desorption of water. The next step is a dehydration of the cellulosic unit between 150-240 °C. Finally, thermal cleavage of the cyclosidic linkage and scission of ether bonds and some C-C bonds via free radical reaction (240-400 °C) and, thereafter, aromatization takes place.

Carbonization: Between 400 and 700 °C, the carbonaceous residue is converted into a graphite-like layer.

Graphitization: Graphitization is carried out under strain at 700-2700 °C to obtain high modulus fiber through longitudinal orientation of the planes.

b- The carbon fiber fabrication from pitch generally consists of the following four steps:

Pitch preparation: It is an adjustment in the molecular weight, viscosity, and crystal orientation for spinning and further heating.

Spinning and drawing: In this stage, pitch is converted into filaments, with some alignment in the crystallites to achieve the directional characteristics.

Stabilization: In this step, some kind of thermosetting to maintain the filament shape during pyrolysis. The stabilization temperature is between 250 and 400 °C.

Carbonization: The carbonization temperature is between 1000-1500 °C.

2.4.2.3. Carbon fibers in meltblown nonwovens

Carbon fibers made from the spinning of molten pitches are of interest because of the high carbon yield from the precursors and the relatively low cost of the starting materials. Stabilization in air and carbonization in nitrogen can follow the formation of melt-blown pitch webs. Processes have been developed with isotropic pitches and with anisotropic mesophase pitches. The mesophase pitch based and melt blown discontinuous carbon fibers have a peculiar structure. These fibers are characterized in that a large number of small domains, each domain having an average equivalent diameter from 0.03 mm to 1mm and a nearly unidirectional orientation of folded carbon layers, assemble to form a mosaic structure on the cross-section of the carbon fibers. The folded carbon layers of each domain are oriented at an angle to the direction of the folded carbon layers of the neighboring domains on the boundary [20].

2.4.2.4. Carbon fibers from isotropic pitch

The isotropic pitch or pitch-like material, i.e., molten polyvinyl chloride, is melt spun at high strain rates to align the molecules parallel to the fiber axis. The thermoplastic fiber is then rapidly cooled and carefully oxidized at a low temperature (<100 °C). The oxidation process is rather slow, to ensure stabilization of the fiber by cross-linking and rendering it infusible. However, upon carbonization, relaxation of the molecules takes place, producing fibers with no significant preferred orientation. This process is not industrially attractive due to the lengthy oxidation step, and only low-quality carbon fibers with no graphitization are produced. These are used as fillers with various plastics as thermal insulation materials [20].

2.4.2.5. Carbon fibers from anisotropic mesophase pitch

High molecular weight aromatic pitches, mainly anisotropic in nature, are referred to as mesophase pitches. The pitch precursor is thermally treated above 350°C to convert it to mesophase pitch, which contains both isotropic and anisotropic phases. Due to the shear stress occurring during spinning, the mesophase molecules orient parallel to the fiber axis. After spinning, the isotropic part of the pitch is made infusible by thermosetting in air at a temperature below it's softening point. The fiber is then carbonized at temperatures up to 1000 °C. The main advantage of this process is that no tension is required during the stabilization or the graphitization, unlike the case of rayon or PANs precursors [21].

2.4.2.6. Structure

The characterization of carbon fiber microstructure has been mainly been performed by x-ray scattering and electron microscopy techniques. In contrast to graphite, the structure of carbon fiber lacks any three dimensional order. In PAN-based fibers, the linear chain structure is transformed to a planar structure during oxidative stabilization and subsequent carbonization. Basal planes oriented along the fiber axis are formed during the carbonization stage. Wide-angle x-ray data suggests an increase in stack height and orientation of basal planes with an increase in heat treatment temperature. A difference in structure between the sheath and the core was noticed in a fully stabilized fiber. The skin has a high axial preferred orientation and thick crystallite stacking. However, the core shows a lower preferred orientation and a lower crystallite height [22].

2.4.2.7. Properties

In general, it is seen that the higher the tensile strength of the precursor the higher is the tenacity of the carbon fiber. Tensile strength and modulus are significantly improved by carbonization under strain when moderate stabilization is used. X-ray and electron diffraction studies have shown that in high modulus type fibers, the crystallites are arranged around the longitudinal axis of the fiber with layer planes highly oriented parallel to the axis. Overall, the strength of a carbon fiber depends on the type of precursor, the processing conditions, heat treatment temperature and the presence of flaws and defects. With PAN based carbon fibers, the strength increases up to a maximum of 1300 ºC and then gradually decreases. The modulus has been shown to increase with increasing temperature. PAN based fibers typically buckle on compression and form kink bands at the innermost surface of the fiber. However, similar high modulus type pitch-based fibers deform by a shear mechanism with kink bands formed at 45° to the fiber axis. Carbon fibers are very brittle. The layers in the fibers are formed by strong covalent bonds. The sheet-like aggregations allow easy crack propagation. On bending, the fiber fails at very low strain [23].

2.4.2.8. Applications

The two main applications of carbon fibers are in specialized technology, which includes aerospace and nuclear engineering, and in general engineering and transportation, which includes engineering components such as bearings, gears, cams, fan blades and automobile bodies. Recently, some new applications of carbon fibers have been found. Others include: decoration in automotive, marine, general aviation interiors, general entertainment and musical instruments and after-market transportation products. Conductivity in electronics technology provides additional new application [24].

The production of highly effective fibrous carbon adsorbents with low diameter, excluding or minimizing external and intra-diffusion resistance to mass transfer, and therefore, exhibiting high sorption rates is a challenging task. These carbon adsorbents can be converted into a wide variety of textile forms and nonwoven materials. Cheaper and newer versions of carbon fibers are being produced from new raw materials. Newer applications are also being developed for protective clothing (used in various chemical industries for work in extremely hostile environments), electromagnetic shielding and various other novel applications. The use of carbon fibers in Nonwovens is in a new possible application for high temperature fire-retardant insulation (eg: furnace material) [25].

2.5. Aramid-definition

Aliphatic polyamides are macromolecules whose structural units are characteristically interlinked by the amide linkage -NH-CO-. The nature of the structural unit constitutes a basis for classification. Aliphatic polyamides with structural units derived predominantly from aliphatic monomers are members of the generic class of nylons, whereas aromatic polyamides in which at least 85% of the amide linkages are directly adjacent to aromatic structures have been designated aramids. The U.S. Federal Trade Commission defines nylon fibers as "a manufactured fiber in which the fiber forming substance is a long chain synthetic polyamide in which less than 85% of the amide linkages (-CO-NH-) are attached directly to two aliphatic groups." Polyamides that contain recurring amide groups as integral parts of the polymer backbone have been classified as condensation polymers regardless of the principal mechanisms entailed in the polymerization process. Though many reactions suitable for polyamide formation are known, commercially important nylons are obtained by processes related to either of two basic approaches: one entails the polycondensation of difunctional monomers utilizing either amino acids or stoichiometric pairs of dicarboxylic acids and diamines, and the other entails the ring-opening polymerization of lactams. The polyamides formed from diacids and diamines are generally described to be of the AABB format, whereas those derived from either amino acids or lactams are of the AB format.

The structure of polyamide fibers is defined by both chemical and physical parameters. The chemical parameters are related mainly to the constitution of the polyamide molecule and are concerned primarily with its monomeric units, end-groups, and molecular weight. The physical parameters are related essentially to chain conformation, orientation of both polymer molecule segments and aggregates, and to crystallinity [26]. This characteristic for single-chain aliphatic polyamides is determined by the structure of the monomeric units and the nature of end groups of the polymer molecules. The most important structural parameter of the noncrystalline (amorphous) phase is the glass transition temperature (T_g) since it has a considerable effect on both processing and properties of the polyamide fibers. It relates to a type of a glass–rubber transition and is defined as the temperature, or temperature range, at which mobility of chain segments or structural units commences. Thus it is a function of the chemical structure; in case of the linear aliphatic polyamides, it is a function of the number of CH_2 units (mean spacing) between the amide groups. As the number of CH_2 unit's increases, T_g decreases. Although T_g is further affected by the nature of the crystalline phase, orientation, and molecular weight, it is associated only with what may be considered the amorphous phase.

Any process affecting this phase exerts a corresponding effect on the glass transition temperature. This is particularly evident in its response to the concentration of water absorbed in polyamides. An increase in water content results in a steady decrease of T_g toward a limiting value. This phenomenon may be explained by a mechanism that entails successive replacement of intercatenary hydrogen bonds in the amorphous phase with water. It may involve a sorption mechanism, according to which 3 mol of water interact with two neighboring amide groups [27].

The properties of aromatic polyamides differ significantly from those of their aliphatic counterparts. This led the U.S. Federal Trade Commission to adopt the term "aramid" to designate fibers of the aromatic polyamide type in which at least 85% of the amide linkages are attached directly to two aromatic rings.

The search for materials with very good thermal properties was the original reason for research into aromatic polyamides. Bond dissociation energies of C-C and C-N bonds in aromatic polyamides are ~20% higher than those in aliphatic polyamides. This is the reason why the decomposition temperature of poly(m-phenylene isophthalamide) MPDI exceeds

450 ºC. Conjugation between the amide group and the aromatic ring in poly(p-phenylene terephthalamide) "PPTA" increases chain rigidity as well as the decomposition temperature, which exceeds 550 ºC.

Obviously, hydrogen bonding and chain rigidity of these polymers translates to very high glass transition temperatures. Using low-molecular-weight polymers, Aharoni [19] measured glass transition temperatures of 272 ºC for MPDI and over 295 ºC for PPTA (which in this case had low crystallinity). Others have reported values as high as 4928 ºC. In most cases the measurement of T_g is difficult because PPTA is essentially 100% crystalline. As one would expect, these values are not strongly dependent on the molecular weight of the polymer above a DP of ~10 [22].

The same structural characteristics that are responsible for the excellent thermal properties of these materials are responsible for their limited solubility as well as good chemical resistance. PPTA is soluble only in strong acids like H_2SO_4, HF, and methanesulfonic acid. Preparation of this polymer via solution polymerization in amide solvents is accompanied by polymer precipitation. As expected, based on its structure, MPDI is easier to solubilize then PPTA. It is soluble in neat amide solvents like N-methyl-2-pyrrolidone (NMP) and dimethylacetamide (DMAc), but adding salts like $CaCl_2$ or LiCl significantly enhances its solubility. The significant rigidity of the PPTA chain (as discussed above) leads to the formation of anisotropic solutions when the solvent is good enough to reach critical minimum solids concentration. The implications of this are discussed in greater detail later in this chapter. It is well known that chemical properties differ significantly between crystalline and noncrystalline materials of the same composition. In general, aramids have very good chemical resistance. Obviously, the amide bond is subject to a hydrolytic attack by acids and bases. Exposure to very strong oxidizing agents results in a significant strength loss of these fibers. In addition to crystallinity, structure consolidation affects the rate of degradation of these materials. The hydrophilicity of the amide group leads to a significant absorption of water by all aramids. While the chemistry is the driving factor, fiber structure also plays a very important role; for example, Kevlar 29 absorbs ~7% water, Kevlar 49~4%, and Kevlar 149 only 1%. Fukuda explored the relationship between fiber crystallinity and equilibrium moisture in great detail. Because of their aromatic character, aramids absorb UV light, which results in an oxidative color change. Substantial exposure can lead to the loss of yarn tensile properties. UV absorption by p-aramids is more pronounced than with m-aramids. In this case a self-screening phenomenon is observed, which makes thin structures more susceptible to degradation than thick ones. Very frequently p-aramids are covered with another material in the final application to protect them. The high degree of aromaticity of these materials also provides significant flame resistance. All commercial aramids have a limited oxygen index in the range of 28-32%, which compares with ~20% for aliphatic polyamides.

Typical properties of commercial aramid fibers are while yarns of m-aramids have tensile properties that are no greater than those of aliphatic polyamides, they do retain useful mechanical properties at significantly higher temperatures. The high glass transition temperature leads to low (less than 1%) shrinkage at temperatures below 250 ºC. In general, mechanical properties of m-aramid fibers are developed on drawing. This process produces fibers with a high degree of morphological homogeneity, which leads to very good fatigue properties. The mechanical properties of p-aramid fibers have been the subject of much study. This is because these fibers were the first examples of organic materials with a very high level of both strength and stiffness. These materials are practical confirmation that nearly perfect orientation and full chain extension are required to achieve mechanical properties approaching those predicted for chemical bonds. In general, the mechanical properties reflect a significant anisotropy of these fibers-covalent bonds in the direction of the fiber axis with hydrogen bonding and van der Waals forces in the lateral direction [26].

Aramid-based reinforcement has been viewed as a more specialty product for applications requiring high modulus and where the potential for electrical conductivity would preclude the use of carbon; for example, aramid sheet is used for all tunnel repairs. Product forms include dry fabrics or unidirectional sheets as well as pre-cured strips or bars. Fabrics or sheets are applied to a concrete surface that has been smoothed (by grinding or blasting) and wetted with a resin (usually epoxy). The composite materials used for concrete infrastructure repair that was initiated in the mid 1980s. After air pockets are removed using rollers or flat, flexible squeegees, a second resin coat might be applied. Reinforcement of concrete structures is important in earthquake prone areas, although steel plate is the primary material used to reinforce and repair concrete structures, higher priced fiber-based sheet structures offer advantages for small sites where ease of handling and corrosion resistance are important. The high strength, modulus, and damage tolerance of aramid-reinforced sheets makes the fiber especially suitable for protecting structures prone to seismic activity. The use of aramid sheet also simplifies the application process. Sheets are light in weight and can be easily handled without heavy machinery and can be applied in confined working spaces. Sheets are also flexible, so surface smoothing and corner rounding of columns are less critical than for carbon fiber sheets [28].

3. All process description

FRP involves two distinct processes, the first is the process whereby the fibrous material is manufactured and formed, and the second is the process whereby fibrous materials are bonded with the matrix during the molding process.

3.1. Fibre process

3.1.1. The manufacture of fibre fabric

Reinforcing Fibre is manufactured in both two dimensional and three dimensional orientations

1. Two Dimensional Fibre Reinforced Polymer are characterized by a laminated structure in which the fibres are only aligned along the plane in x-direction and y-direction of the material. This means that no fibres are aligned in the through thickness or the z-direction, this lack of alignment in the through thickness can create a disadvantage in cost

and processing. Costs and labour increase because conventional processing techniques used to fabricate composites, such as wet hand lay-up, autoclave and resin transfer molding, require a high amount of skilled labour to cut, stack and consolidate into a preformed component.

2. Three-dimensional Fibre Reinforced Polymer composites are materials with three dimensional fibre structures that incorporate fibres in the x-direction, y-direction and z-direction. The development of three-dimensional orientations arose from industry's need to reduce fabrication costs, to increase through-thickness mechanical properties, and to improve impact damage tolerance; all were problems associated with two dimensional fibre reinforced polymers [28].

3.1.2. The manufacture of fibre preforms

Fibre preforms are how the fibres are manufactured before being bonded to the matrix. Fibre preforms are often manufactured in sheets, continuous mats, or as continuous filaments for spray applications. The four major ways to manufacture the fibre preform is though the textile processing techniques of Weaving, knitting, braiding and stitching.

1. Weaving can be done in a conventional manner to produce two-dimensional fibres as well in a multilayer weaving that can create three-dimensional fibres. However, multilayer weaving is required to have multiple layers of warp yarns to create fibres in the z-direction creating a few disadvantages in manufacturing, namely the time to set up all the warp yarns on the loom. Therefore most multilayer weaving is currently used to produce relatively narrow width products or high value products where the cost of the preform production is acceptable. Another Fibre-reinforced plastic 3D one of the main problems facing the use of multilayer woven fabrics is the difficulty in producing a fabric that contains fibres oriented with angles other than 0º and 90º to each other respectively.

2. The second major way of manufacturing fibre preforms is braiding. Braiding is suited to the manufacture of narrow width flat or tubular fabric and is not as capable as weaving in the production of large volumes of wide fabrics. Braiding is done over top of mandrels that vary in cross-sectional shape or dimension along their length. Braiding is limited to objects about a brick in size. Unlike the standard weaving process, braiding can produce fabric that contains fibres at 45 degrees angles to one another. Braiding three-dimensional fibres can be done using four steps, two-step or Multilayer Interlock Braiding. Four step or row and column braiding utilizes a flat bed containing rows and columns of yarn carriers that form the shape of the desired preform. Additional carriers are added to the outside of the array, the precise location and quantity of which depends upon the exact preform shape and structure required. There are four separate sequences of row and column motion, which act to interlock the yarns and produce the braided preform. The yarns are mechanically forced into the structure between each step to consolidate the structure in a similar process to the use of a reed in weaving.Two-step braiding is unlike the four step process because the two-step includes a

large number of yarns fixed in the axial direction and a fewer number of braiding yarns. The process consists of two steps in which the braiding carriers move completely through the structure between the axial carriers. This relatively simple sequence of motions is capable of forming performs of essentially any shape, including circular and hollow shapes. Unlike the four steps process the two steps process does not require mechanical compaction the motions involved in the process allows the braid to be pulled tight by yarn tension alone. The last type of braiding is multi-layer interlocking braiding that consists of a number of standard circular braiders being joined together to form a cylindrical braiding frame. This frame has a number of parallel braiding tracks around the circumference of the cylinder but the mechanism allows the transfer of yarn carriers between adjacent tracks forming a multilayer braided fabric with yarns interlocking to adjacent layers.

The multilayer interlock braid differs from both the four step and two-step braids in that the interlocking yarns are primarily in the plane of the structure and thus do not significantly reduce the in-plane properties of the perform. The four step and two step processes produce a greater degree of interlinking as the braiding yarns travel through the thickness of the preform, but therefore contribute less to the in-plane performance of the preform. A disadvantage of the multilayer interlock equipment is that due to the conventional sinusoidal movement of the yarn carriers to form the preform, the equipment is not able to have the density of yarn carriers that is possible with the two step and four step machines.

3.   Knitting fibre preforms can be done with the traditional methods of Warp and [Weft] Knitting, and the fabric produced is often regarded by many as two-dimensional fabric, but machines with two or more needle beds are capable of producing multilayer fabrics with yams that traverse between the layers. Developments in electronic controls for needle selection and knit loop transfer and in the sophisticated mechanisms that allow specific areas of the fabric to be held and their movement controlled. This has allowed the fabric to form itself into the required three-dimensional perform shape with a minimum of material wastage.

4.   Stitching is arguably the simplest of the four main textile manufacturing techniques and one that can be performed with the smallest investment in specialized machinery. Basically the stitching process consists of inserting a needle, carrying the stitch thread, through a stack of fabric layers to form a 3D structure. The advantages of stitching are that it is possible to stitch both dry and prepreg fabric, although the tackiness of prepare makes the process difficult and generally creates more damage within the prepreg material than in the dry fabric. Stitching also utilizes the standard two-dimensional fabrics that are commonly in use within the composite industry therefore there is a sense of familiarity concerning the material systems. The use of standard fabric also allows a greater degree of flexibility in the fabric lay-up of the component than is possible with the other textile processes, which have restrictions on the fibre orientations that can be produced.

3.1.3. Molding processes

There are two distinct categories of molding processes using FRP plastics; this includes composite molding and wet molding. Composite molding uses Prepreg FRP, meaning the plastics are fibre reinforced before being put through further molding processes. Sheets of Prepreg FRP are heated or compressed in different ways to create geometric shapes. Wet molding combines fibre reinforcement and the matrix or resist during the molding process. The different forms of composite and wet molding, are listed below.

3.2. Composite molding

3.2.1. Bladder molding

Individual sheets of prepreg material are laid -up and placed in a female-style mould along with a balloon-like bladder. The mould is closed and placed in a heated press. Finally, the bladder is pressurized forcing the layers of material against the mould walls. The part is cured and removed from the hot mould. Bladder molding is a closed molding process with a relatively short cure cycle between 15 and 60 minutes making it ideal for making complex hollow geometric shapes at competitive costs.

3.2.2. Compression molding

A "preform" or "charge", of SMC, BMC or sometimes prepreg fabric, is placed into mould cavity. The mould is closed and the material is compacted & cured inside by pressure and heat. Compression molding offers excellent detailing for geometric shapes ranging from pattern and relief detailing to complex curves and creative forms, to precision engineering all within a maximum curing time of 20 minutes.

3.2.3. Autoclave – Vacuum bag

Individual sheets of prepreg material are laid-up and placed in an open mold. The material is covered with release film, bleeder/breather material and a vacuum bag. A vacuum is pulled on part and the entire mould is placed into an autoclave (heated pressure vessel). The part is cured with a continuous vacuum to extract entrapped gasses from laminate. This is a very common process in the aerospace industry because it affords precise control over the molding process due to a long slow cure cycle that is anywhere from one to two hours. This precise control creates the exact laminate geometric forms needed to ensure strength and safety in the aerospace industry, but it is also slow and lab our intensive, meaning costs often confine it to the aerospace industry.

3.2.4. Mandrel wrapping

Sheets of prepreg material are wrapped around a steel or aluminum mandrel. The prepreg material is compacted by nylon or polypropylene cello tape. Parts are typically batch cured by hanging in an oven. After cure the cello and mandrel are removed leaving a hollow carbon tube. This process creates strong and robust hollow carbon tubes.

3.2.5. Wet layup

Fibre reinforcing fabric is placed in an open mould and then saturated with a wet (resin) by pouring it over the fabric and working it into the fabric and mould. The mould is then left so that the resin will cure, usually at room temperature, though heat is sometimes used to ensure a proper curing process. Glass fibres are most commonly used for this process, the results are widely known as fibreglass, and are used to make common products like skis, canoes, kayaks and surf boards.

3.2.6. Chopper gun

Continuous strand of fibreglass are pushed through a hand-held gun that both chops the strands and combines them with a catalyzed resin such as polyester. The impregnated chopped glass is shot onto the mould surface in whatever thickness the design and human operator think is appropriate. This process is good for large production runs at economical cost, but produces geometric shapes with less strength than other molding processes and has poor dimensional tolerance.

3.2.7. Filament winding

Machines pull fibre bundles through a wet bath of resin and wound over a rotating steel mandrel in specific orientations Parts are cured either room temperature or elevated temperatures. Mandrel is extracted, leaving a final geometric shape but can be left in some cases.

3.2.8. Pultrusion

Fibre bundles and slit fabrics are pulled through a wet bath of resin and formed into the rough part shape. Saturated material is extruded from a heated closed die curing while being continuously pulled through die. Some of the end products of the pultrusion process are structural shapes, i.e. beam, angle, channel and flat sheet. These materials can be used to create all sorts of fibreglass structures such as ladders, platforms, handrail systems tank, pipe, and pump supports.

3.3. Resin infusion

Fabrics are placed into a mould which wet resin is then injected into. Resin is typically pressurized and forced into a cavity which is under vacuum in the RTM (Resin Transfer Molding) process. Resin is entirely pulled into cavity under vacuum in the VARTM (Vacuum Assisted Resin Transfer Molding) process. This molding process allows precise tolerances and detailed shaping but can sometimes fail to fully saturate the fabric leading to weak spots in the final shape.

3.3.1. Advantages and limitations

FRP allows the alignment of the glass fibres of thermoplastics to suit specific design programs. Specifying the orientation of reinforcing fibres can increase the strength and resistance to deformation of the polymer. Glass reinforced polymers are strongest and most resistive to deforming forces when the polymers fibres are parallel to the force being exerted, and are weakest when the fibres are perpendicular. Thus this ability is at once both an advantage and a limitation depending on the context of use. Weak spots of perpendicular fibres can be used for natural hinges and connections, but can also lead to material failure when production processes fail to properly orient the fibres parallel to expected forces. When forces are exerted perpendicular to the orientation of fibres the strength and elasticity of the polymer is less than the matrix alone. In cast resin components made of glass reinforced polymers such as UP and EP, the orientation of fibres can be oriented in two-dimensional and three-dimensional weaves. This means that when forces are possibly perpendicular to one orientation, they are parallel to another orientation; this eliminates the potential for weak spots in the polymer.

3.3.2. Failure modes

Structural failure can occur in FRP materials when:

- Tensile forces stretch the matrix more than the fibres, causing the material to shear at the interface between matrix and fibres.

- Tensile forces near the end of the fibres exceed the tolerances of the matrix, separating the fibres from the matrix.

- Tensile forces can also exceed the tolerances of the fibres causing the fibres themselves to fracture leading to material failure [29].

3.3.3. Material requirements

The matrix must also meet certain requirements in order to first be suitable for the FRP process and ensure a successful reinforcement of it. The matrix must be able to properly saturate, and bond with the fibres within a suitable curing period. The matrix should preferably bond chemically with the fibre reinforcement for maximum adhesion. The matrix must also completely envelope the fibres to protect them from cuts and notches that would reduce their strength, and to transfer forces to the fibres. The fibres must also be kept separate from each other so that if failure occurs it is localized as much as possible, and if failure occurs the matrix must also debond from the fibre for similar reasons. Finally the matrix should be of a plastic that remains chemically and physically stable during and after reinforcement and molding processes. To be suitable for reinforcement material fibre additives must increase the tensile strength and modulus of elasticity of the matrix and meet the following conditions; fibres must exceed critical fibre content; the strength and rigidity of fibres itself must exceed the strength and rigidity of the matrix alone; and there must be optimum bonding between fibres and matrix.

3.4. Glass fibre material

FRPs use textile glass fibres; textile fibres are different from other forms of glass fibres used for insulating applications. Textile glass fibres begin as varying combinations of SiO_2, Al_2O_3, B_2O_3, CaO, or MgO in powder form. These mixtures are then heated through a direct melt process to temperatures around 1300 degrees Celsius, after which dies are used to extrude filaments of glass fibre in diameter ranging from 9 to 17 µm. These filaments are then wound into larger threads and spun onto bobbins for transportation and further processing. Glass fibre is by far the most popular means to reinforce plastic and thus enjoys a wealth of production processes, some of which are applicable to aramid and carbon fibres as well owing to their shared fibrous qualities. Roving is a process where filaments are spun into larger diameter threads. These threads are then commonly used for woven reinforcing glass fabrics and mats, and in spray applications. Fibre fabrics are web-form fabric reinforcing material that has both warped and weft directions. Fibre mats are web-form non-woven mats of glass fibres. Mats are manufactured in cut dimensions with chopped fibres, or in continuous mats using continuous fibres. Chopped fibre glass is used in processes where lengths of glass threads are cut between 3 and 26 mm, threads are then used in plastics most commonly intended for moulding processes. Glass fibre short strands are short 0.2–0.3 mm strands of glass fibres that are used to reinforce thermoplastics most commonly for injection moulding.

3.5. Aramid fibre material process

Aramid fibres are most commonly known Kevlar, Nomex and Technora. Aramids are generally prepared by the reaction between an amine group and a carboxylic acid halide group (aramid); commonly this occurs when an aromatic polyamide is spun from a liquid concentration of sulfuric acid into a crystallized fibre. Fibres are then spun into larger threads in order to weave into large ropes or woven fabrics (Aramid) [29]. Aramid fibres are manufactured with varying grades to base on varying qualities for strength and rigidity, so that the material can be somewhat tailored to specific design needs concerns, such as cutting the tough material during manufacture.

3.6. FRP, applications

Fibre-reinforced plastics are best suited for any design program that demands weight savings, precision engineering, finite tolerances, and the simplification of parts in both production and operation. A molded polymer artifact is cheaper, faster, and easier to manufacture than cast aluminum or steel artifact, and maintains similar and sometimes better tolerances and material strengths. The Mitsubishi Lancer Evolution IV also used FRP for its spoiler material [30-32].

3.6.1. Carbon fibre reinforced polymers

Rudder of commercial airplane

• Advantages over a traditional rudder made from sheet aluminum are:

- 25% reduction in weight

- 95% reduction in components by combining parts and forms into simpler molded parts.

- Overall reduction in production and operational costs, economy of parts results in lower production costs and the weight savings create fuel savings that lower the operational costs of flying the airplane.

3.6.2. Structural applications of FRP

FRP can be applied to strengthen the beams, columns and slabs in buildings. It is possible to increase strength of these structural members even after these have been severely damaged due to loading conditions. For strengthening beams, two techniques are adopted. First one is to paste FRP plates to the bottom (generally the tension face) of a beam. This increases the strength of beam, deflection capacity of beam and stiffness (load required to make unit deflection). Alternatively, FRP strips can be pasted in 'U' shape around the sides and bottom of a beam, resulting in higher shear resistance. Columns in building can be wrapped with FRP for achieving higher strength. This is called wrapping of columns. The technique works by restraining the lateral expansion of the column. Slabs may be strengthened by pasting FRP strips at their bottom (tension face). This will result in better performance, since the tensile resistance of slabs is supplemented by the tensile strength of FRP. In the case of beams and slabs, the effectiveness of FRP strengthening depends on the performance of the resin chosen for bonding [32].

3.6.3. Glass fibre reinforced polymers

Engine intake manifolds are made from glass fibre reinforced PA 66.

- Advantages this has over cast aluminum manifolds are:

- Up to a 60% reduction in weight

- Improved surface quality and aerodynamics

- Reduction in components by combining parts and forms into simpler molded shapes. Automotive gas and clutch pedals made from glass fibre reinforced PA 66 (DWP 12-13)

- Advantages over stamped aluminum are:

- Pedals can be molded as single units combining both pedals and mechanical linkages simplifying the production and operation of the design.

- Fibres can be oriented to reinforce against specific stresses, increasing the durability and safety.

3.6.4. Design considerations

FRP is used in designs that require a measure of strength or modulus of elasticity those non-reinforced plastics and other material choices are either ill suited for mechanically or economically. This means that the primary design consideration for using FRP is to ensure that the material is used economically and in a manner that takes advantage of its structural enhancements specifically. This is however not always the case, the orientation of fibres also creates a material weakness perpendicular to the fibres. Thus the use of fibre reinforcement and their orientation affects the strength, rigidity, and elasticity of a final form and hence the operation of the final product itself. Orienting the direction of fibres either, unidirectional, 2-dimensionally, or 3-dimensionally during production affects the degree of strength, flexibility, and elasticity of the final product. Fibres oriented in the direction of forces display greater resistance to distortion from these forces and vice versa, thus areas of a product that must withstand forces will be reinforced with fibres in the same direction, and areas that require flexibility, such as natural hinges, will use fibres in a perpendicular direction to forces. Using more dimensions avoids this either or scenario and creates objects that seek to avoid any specific weak points due to the unidirectional orientation of fibres. The properties of strength, flexibility and elasticity can also be magnified or diminished through the geometric shape and design of the final product. These include such design consideration such as ensuring proper wall thickness and creating multifunctional geometric shapes that can be molding as single pieces, creating shapes that have more material and structural integrity by reducing joints, connections, and hardware [30].

3.6.5. Disposal and recycling concerns

As a subset of plastic FR plastics are liable to a number of the issues and concerns in plastic waste disposal and recycling. Plastics pose a particular challenge in recycling processes because they are derived from polymers and monomers that often cannot be separated and returned to their virgin states, for this reason not all plastics can be recycled for re-use, in fact some estimates claim only 20% to 30% of plastics can be material recycled at all. Fibre reinforced plastics and their matrices share these disposal and environmental concerns. In addition to these concerns, the fact that the fibres themselves are difficult to remove from the matrix and preserve for re-use means FRP amplify these challenges. FRP are inherently difficult to separate into base a material that is into fibre and matrix, and the Fibre-reinforced plastic matrix into separate usable plastic, polymers, and monomers. These are all concerns for environmentally informed design today, but plastics often offer savings in energy and economic savings in comparison to other materials, also with the advent of new more environmentally friendly matrices such as bioplastics and UV-degradable plastics, FRP will similarly gain environmental sensitivity [29].

4. Mechanical properties measurements

4.1. Strength

Strength is a mechanical property that you should be able to relate to, but you might not know exactly what we mean by the word "strong" when are talking about polymers. First, there is more than one kind of strength. There is tensile strength. A polymer has tensile strength if it is strong when one pulls on it. Tensile strength is important for a material that is going to be stretched or under tension. Fibers need good tensile strength.

Then there is compressional strength. A polymer sample has compressional strength if it is strong when one tries to compress it. Concrete is an example of a material with good compressional strength. Anything that has to support weight from underneath has to have good compressional strength [32]. There is also flexural strength. A polymer sample has flexural strength if it is strong when one tries to bend it.

There are other kinds of strength we could talk about. A sample torsional strength if it is strong when one tries to twist it. Then there is impact strength. A sample has impact strength if it is strong when one hits it sharply and suddenly, as with a hammer.

To measure the tensile strength of a polymer sample, we take the sample and we try to stretch. We usually stretch it with a machine for these studies. This machine simply has clamps on each end of the sample, then, when you turn it on it stretches the sample. While it is stretching the sample, it measures the amount of force (F) that it is exerting. When we know the force being exerted on the sample, we then divide that number by the cross-sectional area (A) of our sample. The answer is the stress that our sample is experiencing. Then, using our machine, we continue to increase the amount of force, and stress naturally, on the sample until it breaks. The stress needed to break the sample is the tensile strength of the material. Likewise, one can imagine similar tests for compressional or flexural strength. In all cases, the strength is the stress needed to break the sample. Since tensile stress is the force placed on the sample divided by the cross-sectional area of the sample, tensile stress, and tensile strength as well, are both measured in units of force divided by units of area, usually N/cm2. Stress and strength can also be measured in megapascals (MPa) or gigapascals (GPa). It is easy to convert between the different units, because 1 MPa = 100 N/cm^2, 1 GPa = 100,000 N/cm^2, and of course 1 GPa = 1,000 MPa. Other times, stress and strength are measured in the old English units of pounds per square inch, or psi. If you ever have to convert psi to N/cm^2, the conversion factor is 1 N/cm^2 = 1.45 psi.

4.2. Elongation

But there is more to understanding a polymer's mechanical properties than merely knowing how strong it is. All strength tells us is how much stress is needed to break something. It doesn't tell us anything about what happens to our sample while we're trying to break it. That's where it pays to study the elongation behavior of a polymer sample. Elongation is a type of deformation. Deformation is simply a change in shape that anything undergoes under stress. When we're talking about tensile stress, the sample deforms by stretching, becoming longer. We call this elongation, of course. Usually we talk about percent elongation, which is just the length the polymer sample is after it is stretched (L), divided by the original length of the sample (L_0), and then multiplied by 100.

There are a number of things we measure related to elongation. Which is most important depends on the type of material one is studying. Two important things we measure are ultimate elongation and elastic elongation. Ultimate elongation is important for any kind of material. It is nothing more than the amount you can stretch the sample before it breaks. Elastic elongation is the percent elongation you can reach without permanently deforming your sample. That is, how much can you stretch it, and still have the sample snap back to its original length once you release the stress on it. This is important if your material is an elastomer. Elastomers have to be able to stretch a long distance and still bounce back. Most of them can stretch from 500 to 1000 % elongation and return to their original lengths without any trouble [32].

4.3. Modulus

In the elastomers are need show the high elastic elongation. But for some other types of materials, like plastics, it usually they not stretch or deform so easily. If we want to know how well a material resists deformation, we measure something called modulus. To measure tensile modulus, we do the same thing as we did to measure strength and ultimate elongation. This time we measure the stress we're exerting on the material, just like we did when we were measuring tensile strength. First, is slowly increasing the amount of stress, and then we measure the elongation the sample undergoes at each stress level. We keep doing this until the sample breaks. This plot is called a stress-strain curve. (Strain is any kind of deformation, including elongation. Elongation is the word we use if we're talking specifically about tensile strain.) The height of the curve when the sample breaks is the tensile strength, of course, and the tensile modulus is the slope of this plot. If the slope is steep, the sample has a high tensile modulus, which means it resists deformation. If the slope is gentle, then the sample has a low tensile modulus, which means it is easily deformed. There are times when the stress-strain curve is not nice and straight, like we saw above. The slope isn't constant as stress increases. The slope, that is the modulus, is changing with stress. In a case like this we usually, the initial slope change as the modulus change [32].

In general, fibers have the highest tensile moduli, and elastomers have the lowest, and plastics have tensile moduli somewhere in between fibers and elastomers.

Modulus is measured by calculating stress and dividing by elongation, and would be measured in units of stress divided by units of elongation. But since elongation is dimensionless, it has no units by which we can divide. So modulus is expressed in the same units as strength, such as N/cm^2.

Intrinsic deformation is defined as the materials' true stress-strain response during homogeneous deformation. Since generally strain localization phenomena occur (like necking, shear banding, crazing and cracking), the measurement of the intrinsic materials' response requires a special experimental set-up, such as a video-controlled tensile or a uniaxial compression test. Although details of the intrinsic response differ per material, a general representation of the intrinsic deformation of polymers can be recognized [33], see Figure 1.

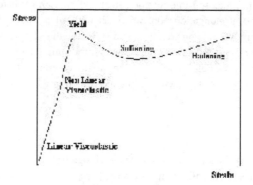

Figure 1. Schematic representation of the intrinsic deformation behavior of a polymer material [33].

4.4. Toughness

That plot of stress versus strain can give us another very valuable piece of information. If one measures the area underneath the stress-strain curve (figure 2), colored red in the graph below, the number you get is something we call toughness.

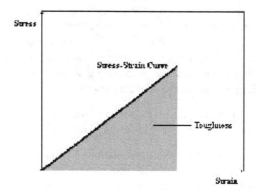

Figure 2. Plot of stress in function of strain.

Toughness is really a measure of the energy a sample can absorb before it breaks. Think about it, if the height of the triangle in the plot is strength, and the base of the triangle is strain, then the area is proportional to strength strain. Since strength is proportional to the force needed to break the sample, and strain is measured in units of distance (the distance the sample is stretched), then strength strain is proportional is force times distance, and as we remember from physics, force times distance is energy.

From a physics point of view the strength, is that strength tells how much force is needed to break a sample, and toughness tells how much energy is needed to break a sample. But that does not really tell you what the practical differences are. What is important knows that just because a material is strong, it isn't necessarily going to be tough as well [34-35].

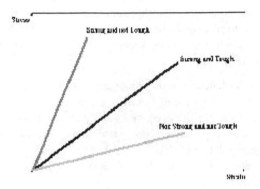

Figure 3. Plot of stress in function of strain, strong and tough concepts.

The gray plot is the stress-strain curve for a sample that is strong, but not tough (figure 3). As you can see, it takes a lot of force to break this sample. Likewise, this sample ca not stretch very much before it breaks. A material like this which is strong, but can not deform very much before it breaks is called brittle [36].

The gray plot is a stress-strain curve for a sample that is both strong and tough. This material is not as strong as the sample in the gray plot, but the area underneath its curve is a lot larger than the area under the gray sample's curve. So it can absorb a lot more energy than the gray-black sample plot.

The gray-black sample elongates a lot more before breaking than the gray sample does. You see, deformation allows a sample to dissipate energy. If a sample can't deform, the energy won't be dissipated, and will cause the sample to break [37].

In real life, we usually want materials to be tough and strong. Ideally, it would be nice to have a material that would not bend or break, but this is the real world. The gray-black sample has a much higher modulus than the red sample. While it is good for materials in a lot of applications to have high moduli and resist deformation, in the real world it is a lot better for a material to bend than to break, and if bending, stretching or deforming in some other way prevents the material from breaking, all the better. So when we design new polymers, or new composites, we often sacrifice a little bit of strength in order to make the material tougher.

4.5. Mechanical properties
of real polymers

The rigid plastics such as polystyrene, poly(methyl methacrylate or polycarbonate can withstand a good deal of stress, but they won't withstand much elongation before breaking. There is not much area under the stress-strain curve at all. So we say that materials like this are strong, but not very tough. Also, the slope of the plot is very steep, which means that it takes a lot of force to deform a rigid plastic. So it is easy to see that rigid plastics have high moduli. In short, rigid plastics tend to be strong, at resist deformation, but they tend not to be very tough, that is, they are brittle.

Flexible plastics like polyethylene and polypropylene are different from rigid plastics in that they don not resist deformation as well, but they tend not to break. The ability to deform is what keeps them from breaking. Initial modulus is high, that is it will resist deformation for awhile, but if enough stress is put on a flexible plastic, it will eventually deform. If you try to stretch it a plastic bag, it will be very hard at first, but once you have stretched it far enough it will give way and stretch easily. The bottom line is that flexible plastics may not be as strong as rigid ones, but they are a lot tougher.

It is possible to alter the stress-strain behavior of a plastic with additives called plasticizers. A plasticizer is a small molecule that makes plastics more flexible. For example, without plasticizers, poly(vinyl chloride), or PVC for short, is a rigid plastic. Rigid unplasticized PVC is used for water pipes. But with plasticizers, PVC can be made flexible enough to use to make things like hoses.

Fibers like Kevlar™, carbon fiber and nylon tend to have stress-strain curves like the aquacolored plot in the graph above. Like the rigid plastics, they are more strong than tough, and do not deform very much under tensile stress. But when strength is what you need, fibers have plenty of it. They are much stronger than plastics, even the rigid ones, and some polymeric fibers, like Kevlar™, carbon fiber and ultra-high molecular weight polyethylene have better tensile strength than steel.

Elastomers like polyisoprene, polybutadiene and polyisobutylene have completely different mechanical behavior from the other types of materials. Take a look at the pink plot in the graph above. Elastomers have very low moduli. You can see this from the very gentle slope of the pink plot, but you probably knew this already. You already knew that it is easy to stretch or bend a piece of rubber [34]. If elastomers did not have low moduli, they would not be very good elastomers.

But it takes more than just low modulus to make a polymer an elastomer. Being easily stretched is not much use unless the material can bounce back to its original size and shape once the stress is released. Rubber bands would be useless if they just stretched and did not bounce back. Of course, elastomers do bounce back, and that is what makes them so amaz ing. They have not just high elongation, but high reversible elongation.

4.6. Tensile properties

The discussion of which types of polymers have which mechanical properties has focused mostly on tensile properties. When we look at other properties, like compressional properties or flexural properties things can be completely different. For example, fibers have very high tensile strength and good flexural strength as well, but they usually have terrible compressional strength. They also only have good tensile strength in the direction of the fibers.

Some polymers are tough, while others are strong, and how one often has to make trade-offs when designing new materials; the design may have to sacrifice strength for toughness, but sometimes we can combine two polymers with different properties to get a new material with some of the properties of both. There are three main ways of doing this, and they are copolymerization, blending, and making composite materials.

The copolymer that combines the properties of two materials is spandex. It is a copolymer containing blocks of elastomeric polyoxyethylene and blocks of a rigid fiber-forming polyurethane. The result is a fiber that stretches. Spandex is used to make stretchy clothing like bicycle pants.

High-impact polystyrene, or HIPS for short, is an immiscible blend that combines the properties of two polymers, styrene and polybutadiene. Polystyrene is a rigid plastic. When mixed with polybutadiene, an elastomer, it forms a phase-separated mixture which has the strength of polystyrene along with toughness supplied by the polybutadiene. For this reason, HIPS is far less brittle than regular polystyrene [38].

In the case of a composite material, we are usually using a fiber to reinforce a thermoset. Thermosets are crosslinked materials whose stress-strain behavior is often similar to plastics. The fiber increases the tensile strength of the composite, while the thermoset gives it compressional strength and toughness.

5. Conclusions

This brief review of FRP has summarized the very broad range of unusual functionalities that these products bring (Polymers, Aramids, Composites, Carbon FRP, and Glass-FRP). While the chemistry plays an important role in defining the scope of applications for which these materials are suited, it is equally important that the final parts are designed to maximize the value of the inherent properties of these materials. Clearly exemplify the constant trade-off between functionality and processability that is an ongoing challenge with these advanced materials. The functionality that allows these materials to perform under extreme conditions has to be balanced against processability that allows them to be economically shaped into useful forms. The ability of a polymer material to deform is determined by the mobility of its molecules, characterized by specific molecular motions and relaxation mechanisms, which are accelerated by temperature and stress. Since these relaxation mechanisms are material specific and depend on the molecular structure, they are used here to establish the desired link with the intrinsic deformation behavior.

Acknowledgement

The author would like to offer a special thanks to Universidad Nacional de San Luis, to Instituto de Física Aplicada, and to Consejo Nacional de Investigaciones Científicas y Técnicas for being generously support used in this research works.

Author details

Martin Alberto Masuelli

Address all correspondence to: masuelli@unsl.edu.ar

Instituto de Física Aplicada, CONICET. Cátedra de Química Física II, Área de Química Física. Facultad de Química, Bioquímica y Farmacia. Universidad Nacional de San Luis. Chacabuco 917 (CP: 5700), San Luis, Argentina

References

[1] http://www.matter.org.uk/matscicdrom/manual/co.html.

[2] Hinton M.J., Soden P.D., Kaddour A.S. Failure Criteria in Fibre-Reinforced-Polymer Composites: The World-Wide Failure Exercise. Elsevier 2004.

[3] Tong L., Mouritz A.P., Bannister M. 3D Fibre Reinforced Polymer Composites. Elsevier 2002.

[4] Ravi Jain, Luke lee. Fiber Reinforced Polymer (FRP) Composites for Infrastructure Applications. Focusing on Innovation, Technology Implementation and Sustainability. Springer 2012.

[5] Kimg Hwee TAN. Fibre Reinforced Polymer. Reinforcement for Concrete Structures. Proceedings of the Sixth International Symposium on FRP Concrete Structures, volume 1-2 (FRPRCS-6). World Scientific 2003.

[6] Erki M.A., and Rizkalla S.H. FRP Reinforcement for Concrete Structures. Concrete International (1993) 48-53.

[7] Han E.H. Meijer, Govaert Leon E. Mechanical performance of polymer systems: The relation between structure and properties. Prog. Polym. Sci. 30 (2005) 915-938.

[8] McGraw-Hill Science & Technology Encyclopedia

[9] Entsiklopediia polimerov, vols. 1–3. Moscow, 1972 77. The Great Soviet Encyclopedia, 3rd Edition (1970-1979). 2010 The Gale Group, Inc.

[10] Giancaspro James, Papakonstantinou Christos, Balaguru P. Mechanical behavior of fire-resistant biocomposite. Composites: Part B 40 (2009) 206-211.

[11] Aubourg, P.F., Crall C., Hadley J., Kaverman R.D., and Miller D.M. Glass Fibers, Ceramics and Glasses. Engineered Materials Handbook, Vol. 4. ASM International, 1991, pp. 1027-31.

[12] McLellan, G.W. and Shand E.B. Glass Engineering Handbook. McGraw-Hill, 1984.

[13] Pfaender, H.G. Schott Guide To Glass. Van Nostrand Reinhold Company, 1983.

[14] Tooley, F.V. "Fiberglass, Ceramics and Glasses", in Engineered Materials 14- 14-14- Handbook, Vol. 4. ASM International, 1991, pp. 402-08.

[15] Hnat, J.G. "Recycling of Insulation Fiberglass Waste". Glass Production Technology International, Sterling Publications Ltd., pp. 81-84.

[16] Webb, R.O. "Major Forces Impacting the Fiberglass Insulation Industry in the 1990s". Ceramic Engineering and Science Proceedings, 1991, pp. 426-31.

[17] How fiberglass is made - material, used, processing, components, dimensions, composition, product, industry, machine, Raw Materials, The Manufacturing Process of fiberglass, Quality Control http://www.madehow.com/Volume-2/Fiberglass.html

[18] Carbon fibers Seen as Having Big Long Term Growth Infrastructure is Next Big Trend Driver, "Advanced Materials & Composites" News, No. 3, 1999.

[19] Structures, Advanced Materials & Composites News, No. 2, 1999.

[20] New Company Launches Carbon Fiber Fabrics for Decorative Applications, Advanced Materials & Composites News, No. 8, 1998

[21] Carbon fibers Electrical Conductivity Found to Offer New Uses, Composites News, No. 3, 1998, 9.

[22] Donnet Jean-Baptiste, Roop Chand Bansal, "Carbon Fibers", published by Marcel Dekker Inc., 1990, p370.

[23] Gullapalli Sravani; Wong Michael S. (2011). Nanotechnology: A Guide to Nano-Objects. Chemical Engineering Progress: 28–32.

[24] Zhao Z., and Gou J. (2009). Improved fire retardancy of thermoset composites modified with carbon nanofibers. Sci. Technol. Adv. Mater. 10: 015005. doi: 10.1088/1468-6996/10/1/015005.

[25] Japan Carbon Fiber Manufacturers Association (English) (http://www.carbonfiber.gr.jp/ english/)

[26] Handbook of Fiber Chemistry. International Fiber Science and Technology Series. Editor Menachem Lewin. Third Edition 2007 CRC Press, Taylor & Francis Group.

[27] Jassal M., Ghosh S. Aramid Fibres – An Overview. Indian Journal of Fibre & Textile Research 27 (2002) 290-306.

[28] José M. García, Félix C. García, Felipe Serna, José L. de la Peña. High-performance aromatic polyamides. Progress in Polymer Science 35 (2010) 623-686.

[29] Smallman, R.E., and Bishop R.J.. Modern Physical Metallurgy and Materials Engineering. 6th ed. Oxford. Butterworth-Heinemann, 1999.

[30] Erhard, Gunter. Designing with Plastics. Trans. Martin Thompson. Munich: Hanser Publishers, 2006.

[31] Rosato Donald V., Rosato Dominick V., Murphy John. Reinforced Plastics Handbook. Elsevier; 2004; page 586

[32] Composite moulding (http:/ / web. archive. org/ web/ 20080215010801/ http:/ / www. quatrocomposites. com/ comp101proc. htm).

[33] Klompen, Edwin T.J. Mechanical properties of solid polymers. Constitutive modelling of long and short term behaviour. Eindhoven: Technische Universiteit Eindhoven, 2005.

[34] FRP Library (http://www.compositepedia.com)

[35] Odian, George. Principles of Polymerization. 3rd ed., J. Wiley, New York, 1991.

[36] Jang, B.Z.; Advanced Polymer Composites: Principles and Applications, ASM International, Materials Park, OH, 1994.

[37] Bhowmick, Anil. Mechanical Properties of Polymers. Volume 1, Material Sciences Engineering. Enciclopedia of Life Support Systems – UNESCO, 2011.

[38] Hashim M.H.M., Sam A.R.M., Hussin M.W. The Future of External Application of Fibre Reinforced Polymer in Civil Infrastructure for Tropical Climates Region. International Journal of Mechanical and Materials Engineering (IJMME), 6, 2 (2011) 147-159.

Chapter 2

Natural Fibre Bio-Composites Incorporating Poly(Lactic Acid)

Eustathios Petinakis, Long Yu, George Simon and Katherine Dean

Additional information is available at the end of the chapter

1. Introduction

In recent years increasing awareness in relation to the worlds petrochemical resources no longer being finite, the rising cost of oil and concerns surrounding climate change and the necessity for reducing our carbon footprint, are resulting in a renewed demand and expediation of development of polymeric materials that are produced from sustainable and ecologically sound raw material feedstocks that are not petrochemically derived and are generally more abundant. Poly(lactic acid) (PLA), being a compostable synthetic polymer produced using monomer feedstock derived from corn starch, satisfies many of the environmental impact criteria required for an acceptable replacement for oil-derived plastics [1]. PLA exhibits mechanical properties that make it useful for a wide range of applications, but mainly in applications that do not require high performance including plastic bags, packaging for food, disposable cutlery and cups, slow release membranes for drug delivery and liquid barrier layers in disposable nappies [2]. However, the wider uptake of PLA is restricted by performance deficiencies, such as its relatively poor impact properties which arise from its inherent brittleness, but also the limited supply and higher cost of PLA compared with commodity polymers such as polyethylene and polypropylene [3].

Polymer composite materials often possess mechanical and physical properties that make them better suited for a wide range of applications than the individual composite components. The use of natural fibers to produce polymer composites having improved mechanical and impact performance is well-documented, and is of particular interest for enhancing the properties of biodegradable polymeric materials such as PLA [4-6]. The benefits of using natural fibers compared with other potential reinforcing agents, such as glass fibers, talc, or carbon fibers, for improving the performance of biodegradable polymers include the retention of the biodegradability of the composite, but generally also exhibit lower density, superior performance, and lower cost due to the abundance of many natural fibres. An example of such a product is wood-flour, a bio-waste product produced from the preparation of timber for the building and other related industries.

PLA/natural fibre composites containing less than 30%w/w fiber have been shown to have increased tensile modulus and reduced tensile strength compared with PLA, and this has been attributed to factors that include the weak interfacial interaction between the hydrophobic PLA matrix and the hydrophilic cellulose fibers, and lack of fiber dispersion due to a high degree of fiber agglomeration. Various methods of modifying the surface of the cellulosic fibers have been explored in an effort to improve the interaction that occurs at the interface between the PLA matrix and the fibers, including esterification [7], acetylation [8], and cyanoethylation [9]. It is apparent that the stronger the molecular interactions that occur at the interface between cellulose fibers and polymer matrix, the resulting interfacial adhesion is stronger and the optimal stress transfer efficiency. The use of coupling agents or compatibilisers have proven to be a much more efficient means of improving interfacial interactions between polymer matrices and cellulose fibers. The strongest adhesion can be achieved when covalent bonds are formed at the interface between cellulose fibers and coupling agent as well as molecular entanglement between coupling agent and the polymer itself [10]. Coupling agents also have the benefits of improving fibre dispersion, since they can also induce better flow of the molten polymer during processing, improve melt elasticity and melt strength in the resulting polymer composites [11].

The nature of the interface/interphase in polymer composites incorporating natural fibers is still not well understood, since many of the chemical reagents used in surface modification of natural fibres and/or coupling agents used do not form covalent bonds, but in most cases are rarely created [12]. Therefore, the aims of this book chapter research will attempt to address this issue and provide a fundamental understanding of the surface and adhesion properties of polymer composite systems incorporating natural fibres. The authors of this book chapter will demonstrate such understanding thru their research into PLA based bio-composites incorporating wood-flour as the natural fiber/filler. It has been demonstrated in the preliminary research that the best path forward for developing polymer composites with enhanced physical properties is by modifying the surface of wood-flour particles to induce chemical bonding at the interface and to enhance compatibilisation with the PLA matrix.

2. Poly(lactic acid) bio-composites based on natural fibres

2.1. Structure and properties of natural fibres

Natural fibres can be classified into five major types: bast, leaf, seed, fruit and wood, depending upon the source. In order to develop polymer composites from natural resources it is important to understand the microstructure and chemical composition of natural fibres. Natural fibres comprise of three principal components: cellulose, lignin and hemicellulose. These three hydroxyl- containing natural polymers are distributed throughout the cell wall.

Cellulose and hemicellulose are polysaccharides. Cellulose is a highly crystalline polymer with a regular structure, which comprises of thousands of anhydroglucose units with a DP (degree of polymerisation) around 10,000 [13]. Cellulose is the major component which is responsible for the inherent strength and stability of the natural fibre. Hemicellulose is a shorter branched polymer composed of various five- and six- carbon ring sugars. The molecular weight is much lower than cellulose but still hemicellulose still contributes to the structure of natural fibres. Lignin is an amorphous, cross-linked polymer network, which consists of an irregular array of variously bonded hydroxy-and methoxy- substituted phenyl propane units. The chemical structure of lignin depends on the source of the wood. Lignin is not as polar as cellulose and the major function of lignin is to function as a chemical adhesive between cellulose fibers.

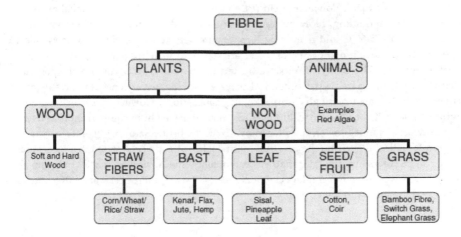

Figure 1. Classification of natural fibres (courtesy of Mohanty et al [14])

Natural fibres also consist of varying amounts pectin, wax and other low molecular weight compounds or extractives. Extractives are described as non-structural components in wood, which are composed of extra cellular and low molecular weight compounds. There are three types of lipophilic extractive compounds: terpenes (and terpenoids), aliphatics (fatty acids and their esters) and phenolic compounds [15]. Aliphatic compounds include alkanes, fatty alcohols, fatty acids, fat esters and waxes. Terponoids include turpentine and resin acids. Phenolic compounds include tannins, flavnoids, lignans, stilbines and tropolones. Extractives can diffuse to the surface of natural fibres during drying, which can influence the degree of adhesion. This is an important factor to consider during the processing of polymer composites incorporating lignocellulosic fillers, since extractives may influence the degree of interfacial adhesion between polymer matrix and lignocellulosic filler. Extractives can be typically removed by solvent extraction or steam distillation or even water treatment if compounds are water soluble. Steam distillation can be used to remove the volatile terpenes, whereas solvent extraction can remove resin acids, fatty alcohols, fatty acids and waxes.

2.2. Structure and properties of PLA

Polymers from renewable resources can be classified into three major groups: natural polymers such as starch and cellulose; synthetic polymers from natural monomers such as polylactic acid (PLA); polymers from microbial fermentation such as polyhydroxybutyrate (PHB). Polylactic acid is one of the most promising biodegradable polymers, which can be derived from natural feedstocks such as corn starch but can also be derived from rice, potatoes, sugar beet and other agricultural waste. Intially, PLA synthesis involves conversion of the raw material feedstock into dextrose, which then undergoes conversion into lactic acid or lactide via a fermentation process in the presence of a catalyst. The lactide undergoes further processing in order to purify the monomer and this is followed by conversion of the purified monomer in into a polymeric form of PLA through polymerisation in the presence of a suitable catalyst [16]. Polylactic Acid (PLA) can be processed by conventional methods such as injection moulding, blow moulding, extrusion and film forming operations, since PLA has a T_g of 55-65°C and a melting temperature between 150-175°C. The mechanical properties of PLA are similar or in most cases are superior too petrochemical polymers, such as polypropylene. Therefore, PLA has attracted great interest as a commodity polymer which is capable of replacing petrochemical polymers like polypropylene and polyethylene, particularly in the area of single use packaging applications. However, PLA exhibits low toughness due to its brittle nature, but also the molecular weight in comparison to conventional polymers, is much lower. In order to overcome the brittle nature of PLA it is useful to incorporate natural fillers into the polymer matrix. It has already been stated that incorporation of natural fillers into polymer matrices can optimise mechanical properties but from an economical viewpoint, natural fillers can make the composites more cost competitive due to their high abundance and lower cost.

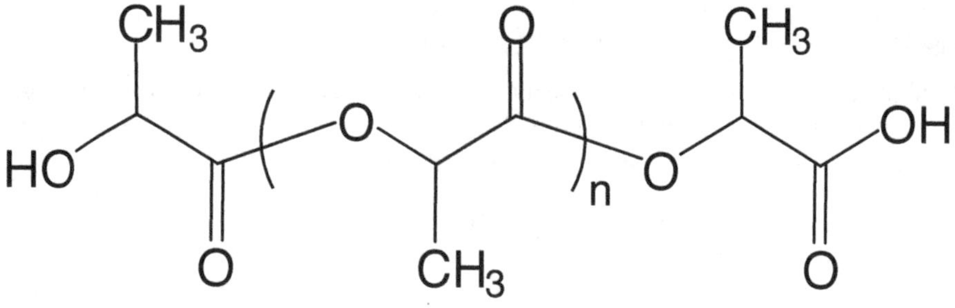

Figure 2. Structure of Poly(lactic acid)

2.3. Natural fibres used in PLA based bio-composites

The major factors that can influence the development of polymer composites using natural fibers are listed as follows [17]:

1. Thermal stability

2. Moisture Content

3. Processability

4. Fibre dispersion in polymer matrix

5. Fibre-matrix adhesion

6. Surface modification of natural fibers

7. Fiber aspect ratio

Oksman et. al. incorporated cellulose fibres as reinforcement in PLA [18]. Due to the brittle nature of PLA, triacetin was used as a plasticizer for the matrix as well as PLA/flax composites in order to improve the impact properties. Plasticizers can be used during processing in order to lower the viscosity of the matrix polymer, which can then facilitate better fiber dispersion within the matrix polymer. Fiber dispersion is a critical factor to be considered during the development of biodegradable natural fibre composites. Shibata et. al. evaluated the use of short abaca fibres in the development of biocomposites using biodegradable polyesters. In this study it was shown that strength and modulus increase with decreasing fibre diameter for both untreated and treated abaca fibre[19].

Wollerdorfer et. al. investigated the influence of plant fibres such as flax, jute, ramie, oil palm fibres and fibres made from regenerated cellulose on the mechanical properties of biodegradable polymers. The so-called biocomposites produced by embedding natural fibres, e.g. flax, hemp, ramie, etc. into a biopolymeric matrix made of derivatives from cellulose, starch, lactic acid, etc., new fibre reinforced materials were developed [20]. Huda et. al. evaluated the use of recycled cellulose in the development of "green" composites using PLA as the matrix and recycled cellulose from newsprint. The physico-mechanical properties of the composites as well as the morphological properties were investigated as a function of varying amounts of recycled cellulose [21]. Bax et al [22] investigated the impact of cordenka and flax fibres on the impact and tensile properties. The study showed that PLA composites with cordenka fibres with a maximum fibre loading of 30% show promise as alternative biocomposites for industrial applications due to optimisation in impact properties. However, both biocomposites showed evidence of poor interfacial adhesion between the PLA matrix and the cordenka and flax fibres, respectively.

Mathew et. al. conducted a study towards developing PLA based high performance nanocomposites using microcrystalline cellulose as reinforcement. The study was concerned with achieving the best possible outcome for dispersion of the MCC within PLA during processing. Comparisons were also made with using wood flour and wood pulp as an alternative reinforcement for PLA [23]. Tzerki et. al. investigated the usefulness of lignocellulosic waste flours derived from spruce, olive husks and paper flours as potential reinforcements for the preparation of cost-effective bio-composites using PLA as the matrix [24]. Petinakis et al studied the effect of wood-flour content on the mechanical properties and fracture behaviour of PLA/wood-flour composites. The results indicated that enhancements in tensile modulus could be achieved, but the interfacial adhesion was poor [25]. Therefore, it can be seen that incorporation of lignocellulosic materials into biodegradable polymer matrices, such as PLA, has the affect of improving mechanical properties, such as tensile modulus. But the strength and toughness of these bio-composites are not necessarily improved. This can be attributed to several reasons, such as the hydrophilic nature of natural fillers, compatibility with the hydrophobic polymer matrix can be problematical. In addition to the poor interaction between the phases, the hydrophilic nature of natural fibres leads to a tendency for fibres to mingle or form agglomerations, which can generally result in low impact properties, especially at high fibre loadings.

In order to overcome these shortcomings a variety of chemical and physical treatments can be utilised to improve fibre-matrix adhesion in biodegradable polymer composites as well as improve dispersion of natural fibres within biopolymer matrices. There are many articles in the public domain that have reported the use of coupling agents and compatibilisers for improving fibre-matrix interfacial adhesion in polymer composite systems incorporating a polyolefinic matrix, such as polypropylene and polyethylene.

3. Strategies for improving interfacial adhesion in PLA/natural fibre composites

Surface modification of natural fillers can be classified into two major types; chemical and physical methods. Surface modification is a critical processing step in the development of biopolymer composites, since natural fillers tend to be highly hydrophilic in nature and in order to improve the compatibilsation with the hydrophobic polymer matrix this level of processing is required. The use of surface modification techniques can facilitate fibre dispersion within polymer matrix as well as improve the fibre-matrix interaction [26]. Some of the techniques that have been previously reported in the literature for improving fibre-matrix adhesion include: treatment of fibres by bleaching, acetylation, esterification, grafting of monomers and the use of bi-functional molecules [27]. The use of coupling agents and compatibilisers has also been widely reported in the development of conventional polymer composites. Coupling agents include silanes, isocyanates, zirconates, titanates and chitosan [28]. One of the most widely reported compatibilisers in the literature has been the use of functional polyolefins such as maleated polypropylene (MAPP) [29-34]. More recently, Xu et al synthesised a novel graft copolymer, polylactide-graft-glycidyl methacrylate (PLA-GMA), which was produced by grafting glycidyl methacrylate onto the PLA chain via free radical polymerisation, which was then used to produce biocomposites using PLA and bamboo flour [35]. All techniques have proven successful in improving the fibre-matrix interactions, which have resulted in polymer composites with greatly improved mechanical properties.

3.1. Chemical techniques

3.1.1. Alkaline treatment

Alkaline treatment is one of the most widely used chemical treatments for natural fibres for use in natural fibre composites. The effect of alkaline treatment on natural fibres is it disrupts the incidence of hydrogen bonding in the network structure, giving rise to additional sites for mechanical interlocking, hence promoting surface roughness and increasing matrix/fibre interpenetration at the interface. During alkaline treatment of lignocellulosic materials, the alkaline treatment removes a degree of the lignin, wax and oils which are present, from the external surface of the fibre cell wall, as well as chain scission of the polymer backbone resulting in short length crystallites. The treatment exposes the hydroxyl groups in the cellulose component to the alkoxide.

In alkaline treatment, wood fibres/flour is immersed in a solution of sodium hydroxide for a given period of time. Beg et. al. studied the effect of the pre-treatment of radiate pine fibre with NaOH and coupling with MAPP in wood fibre reinforced polypropylene composites. It was found that fibre pre-treatment with NaOH resulted in an improvement in the stiffness of the composites (at 60% fibre loading) as a function NaOH concentration, however at the same time, a decrease was observed in the strength of the composite [36]. The reason for a reduction in the tensile strength was attributed to a weakening of the cohesive strength of the fibre, as a result of alkali treatment. The use of alkali treatment in conjunction with MAPP was found to improve the fibre/matrix adhesion. However, it seems that only small concentrations of NaOH can be used to treat fibres, otherwise the cohesive strength can be compromised. Ichazo et. al also studied the addition of alkaline treated wood flour in polypropylene/wood flour composites. It was shown that alkaline treatment only improved fibre dispersion within the polypropylene matrix but not the fibre-matrix adhesion. This was attributed to a greater concentration of hydroxyl groups present, which increased the hydrophilic nature of the composites. As a result, no significant improvement was observed in the mechanical properties of the composites and a reduction in the impact properties [37]. From previous studies it is shown that the optimal treatment conditions for alkalization must be investigated further in order to improve mechanical properties. Care must be taken in selecting the appropriate concentration, treatment time and temperature, since at certain conditions the tensile properties are severely compromised. Islam et al studied the effect of alkali treatment on hemp fibres, which were utilised to produce PLA biocomposites incorporating hemp fibres. This study showed that crystallinity in PLA was increased due to the nucleation of hemp fibres following alkaline treatment. The degree of crystallinity had a positive impact on the mechanical and impact performance of the resulting composites with alkaline treated hemp fibres as opposed to the composites without treated hemp fibres.

3.1.2. Silane treatment

Silane coupling agents have been used traditionally in the past in the development of conventional polymer composites reinforced with glass fibres. Silane is a class of silicon hydride with a chemical formula SiH_4. Silane coupling agents have the potential to reduce the incidence of hydroxyl groups in the fibre-matrix interface. In the presence of moisture, hydrolysable alkoxy groups result in the formation of silanols. Silanols react with hydroxyl groups of the fibre, forming a stable, covalently bonded structure with the cell wall. As a result, the hydrocarbon chains provided by the reaction of the silane produce a cross-linked network due to covalent bonding between fibre and polymer matrix. This results in a hydrophobic surface in the fibre, which in turn increases the compatibility with the polymer matrix. As mentioned earlier silane coupling agents have been effective for the treatment of glass fibres for the reinforcement of polypropylene. Silane coupling agents have also been found to be useful for the pre-treatment of natural fibres in the development of polymer composites. Wu et. al. demonstrated that wood fibre/polypropylene composites containing fibres pre-treated with a vinyl-tri methoxy silane significantly improved the tensile properties. It was discovered that the significant improvement in tensile properties was directly related to a strong interfacial bond caused by the acid/water condition used in the fibre pre-treatment [38].

In a study by Bengtsson et al. the use of silane technology in crosslinking polyethylene-wood flour composites was investigated [39]. Composites of polyethylene with wood-flour were reacted in-situ with silanes using a twin screw extruder. The composites showed improvements in toughness and creep properties and the likely explanation for this improvement was that part of the silane was grafted onto polyethylene and wood, which resulted in a cross-linked network structure in the polymer with chemical bonds occurring at the surface of wood. X-ray microanalysis showed that most of the silane was found within close proximity to the wood-flour. It is known that silanes can interact with cellulose through either free radical or condensation reaction but also through covalent bonding by the reaction of silanol groups and free hydroxyl groups at the surface of wood, however the exact mechanism could not be ascertained. In a study by González et al, focused on the development of PLA based composites incorporating untreated and silane treated sisal and kraft cellulose fibres [40]. The tensile properties of the resulting composites did not present any major statistical difference between composites with untreated cellulose fibres and silane treated cellulose fibres, which suggested that silane treatment of the cellulose fibres did not contribute to further optimisation in the reinforcing affect of the cellulose fibres. The analysis of the high resolution C1s spectra (XPS) indicates that for C_1 (C-C, C-H), the percentage of lignin in the intreated sisal fibres was higher, in comparison with kraft fibres. But after modification with silanes, the C_1 signal decreases for sisal fibres, which shows that attempted grafting with the silane has resulted in removal of lignin and exposed further cellulose. The higher C_1 signal reported for kraft fibres suggested some grafting with silane as a result of the contribution from the alkyl chain of the attached silanol, but no further characterisation was provided to support grafting of silanes to kraft fibres.

3.1.3. Esterification of natural fibres

This section reviews research into the modification of wood constituents with organic acid anhydrides. Anhydrides can be classified into two major groups: non-cyclic anhydrides (i.e. Acetic) and cyclic anhydrides (i.e. Maleic). Of the non-cyclic anhydrides, Acetylation with Acetic Anhydride is the most widely reported [41-43]. The reaction involves the conversion of a hydroxyl group into an ester group by the chemical affiliation of the carboxylic group of the anhydride with the free hydroxyl groups in cellulose. Reactions involving non-cyclic anhydrides are quite cumbersome as there are several steps involved during the treatment. These reactions also require the use of strong bases or catalysts to facilitate the reaction. Although the use of non-cyclic anhydrides can generally lead to good yields a large proportion of the treated cellulose can contain free anhydride, which cannot be easily removed from the treated cellulose. Generally, the modified cellulose may comprise of a distinct odour, which suggests the presence of free anhydride. The other drawback of the use of non-cyclic anhydrides is the formation of acid by-products, which are generally present in the modified cellulose. Pyridine, a catalyst used in the reaction, acts by swelling the wood and extracting lignin to expose the cellular structure of the cellulose. This facilitates the exposure of the free hydroxyl groups in cellulose to the anhydride. However, due to the aggressive nature of pyridine, it can also degrade and weaken the structure of the cell wall, which may not allow efficient modification. The effect of esterification on natural fibres is it imparts hydrophobicity, which makes them more compatible with the polymer matrix.

Tserki et. al. investigated the reinforcing effect of lignocellulosic fibres, incorporating flax, hemp and wood, on the mechanical properties of Bionolle, an aliphatic polyester [44]. The use of acetic anhydride treatment of the fibres was proven not to be as effective for improving the matrix tensile strength, compared with other techniques such as compatibilisation; however it did reduce the water absorption of the fibres. Lower tensile strengths were reported for composites reinforced with wood fibre, compared with flax and hemp. This may be attributed to the nature of the fibres, since flax and hemp are fibrous, whereas wood fibre is more flake like in nature with an irregular size and shape. The type and nature of lignocellulose fibres (chemical composition and structure) is of paramount importance in the development of polymer composites. It is shown that different fibres behave differently after various treatments. On the other hand, reactions of cellulose with cyclic anhydrides have also been performed [45]. Reactions involving cyclic anhydrides generally do not result in the formation of by-products and reactions can be performed with milder solvents, which don't interfere with the cell wall structure of cellulose. In order to facilitate reactions of wood flour with cyclic anhydrides it is important that the wood flour be pre-treated. Pre-treatment requires immersion of the wood flour in a suitable solvent, such as NaOH. This process is otherwise known as Mercerization, which is thought to optimise fiber-surface characteristics, by removing natural impurities such as pectin, waxy substances and natural oils. It is widely reported that the wood alone does not readily react with esterifying agents, since the hydroxyl groups required for reaction are usually masked by the presence of these natural impurities.

3.1.4. Isocyanate treatment

Isocyanates are compounds containing the isocyanate functional group $-N=C=O$, which is highly reactive with hydroxyl groups in lignocellulose materials. The general reaction for cellulose with an isocyanate coupling agent is shown Equation 1:

Equation 1. Possible reaction mechanism of MDI with wood-flour

R can represent any chemical group, such as alkyl or phenyl. Pickering et al studied the effects of Poly[methylene(polyphenyl isocyanate) and maleated coupling agents on New Zealand radiata pine fibre-polypropylene composites. A modest improvement in strength (4%) was reported with the addition of isocyanate to the polymer matrix over the matrix alone. When the radiata pine was treated with isocyanate and added to matrix, the strength improved by 11.5% over untreated radiata pine, and the modulus exhibited a significant improvement of 77% [46]. It appears that lignin content in wood fibres plays a significant role in relation to the ability of certain functional groups to interact with the cellulose component. The modest gains in tensile strength with the isocyanate can be attributed to the greater percentage of lignin in radiata pine.

X-ray mapping using Electron Probe Microanalysis presents a useful technique for evaluating the extent of cross-linking with MDI in biopolymer composites. Analysis of polished cross-sections was performed on unmodified wood-flour composite and the composites with MDI-mediated wood-flour. The aim of this was to detect the presence of nitrogen in the composites, which would indicate the extent of cross-linking in the modified PLA/wood-flour composites with MDI. The micrographs with the X-ray mapping of micro-composite with unmodified wood-flour composite and (b) micro-composite with MDI-modified wood-flour are shown in Figure 3. The nitrogen in the composite is depicted in the micrograph by the regions colored in green. The composite with unmodified wood-flour (Figure 3(a)) shows some nitrogen but this was expected since wood in its native form comprises traceable amounts (<0.75%) of nitrogen. Figure 3(b) depicts the composite with MDI-modified wood-flour and shows a greater concentration of nitrogen, presumably associated with MDI, in close proximity to the particle and the fibre lumen (cells) of the wood-flour particles and some concentrated areas at the interface. Similar observations were also reported by Bengtsson et al., which demonstrated the X-ray mapping of silicon from the silane used to modify wood-flour for polyethylene composites[47]. This suggests some reaction of the MDI-modified wood-flour with the PLA matrix creating a cross-linked structure with chemical bonds joining MDI-modified wood-flour with the PLA matrix. This provides further evidence of the improvement in the mechanical properties as a result of an improvement in the interfacial interaction between PLA and the wood-flour particles. MDI appears to be also spread throughout the PLA matrix suggesting that part of it remains un-reacted within the host polymer.

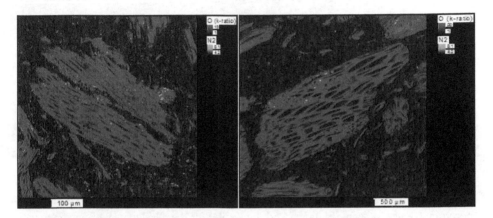

Figure 3. Electron Probe Microanalysis of PLA/wood-flour micro-composites containing (a) unmodified wood-flour (b) MDI-mediated wood-flour (wood-flour content = 30% w/w)

Ecotoxicity is an important factor to consider when developing polymer composites from renewable resources. Isocyanate compounds, such as MDI, may not be regarded as a viable treatment method for natural fibres. Isocyanates upon decomposition in water can result in the formation of diamines. The decomposition products, such as 4, 4′-methylenedianiline and 2,4-diaminotoluene are suspected to be cancer causing agents and may also cause hepatitis in humans [48]. An alternative isocyanate that has been reported in the literature is lysine-based di-isocyanate. Lee et al reported the use of LDI as a coupling agent in the development of biodegradable polymers produced from poly lactic acid/bamboo fibre and poly(butylene succinate)/bamboo fibre. LDI is based upon Lysine, a naturally occurring amino acid with two amino groups and one carboxylic group. LDI can react with hydroxyl groups in cellulose, forming an isocyanate bridge, which can then readily react with the carboxylic and hydroxyl groups of the matrix polymer. MDI has previously been reported in the compatibilization of PLA and starch blends [49]. Wang discovered that blends of PLA with 45% wheat starch and 0.5% MDI resulted in composites with the highest tensile strength. It was also shown that moisture absorption increased as a function of increasing starch content. Water absorption can influence the mechanical properties of the composite. The moisture in the composite can react with MDI, which can effect interfacial interaction between starch mediated MDI with the PLA matrix by reducing the tensile strength or having a limited improvement. The reaction of moisture with MDI has also been reported in another paper [50] by Yu et al. It was interesting to note that the highest strength was achieved at 45%. This can be attributed to two major reasons: the level of water in the blend can aid processing of the PLA, whereby the water behaves as a plasticiser, and secondly, the viscosity of the PLA at this level of water content maybe just sufficient to allow optimum dispersion of the starch particles within the PLA matrix. However, in order to utilise these materials in commercial applications such as for short term packaging these materials would require water proofing on the surface in order to prevent the rapid degradation.

3.2. Physical techniques

Physical methods [51-54] reported in the literature are the use of corona or plasma treaters for modifying cellulose fibres for conventional polymers. In recent years the use of plasma for treatment of natural fibres has gained more prominence as this provides a more "greener" alternative for the treatment of natural fibres for the development of polymer composites, but is of particular interest to polymer composites incorporating biopolymer matrices, since this technique provides further credence to the whole idea of "green materials. Sustainability and end of life after use are important considerations to make when developing polymer composites from renewable resources is the toxicity and environmental impact of using various chemical or physical methods for improving the properties of these materials. Some chemical techniques may be toxic, e.g. isocyanates are carcinogenic, and therefore, the use of such agents may not be feasible for the development of polymer composites from renewable resources. Physical methods involving plasma treatments have the ability to change the surface properties of natural fibres by formation of free radical species (ions, electrons) on the surfaces of natural fibres [55]. During plasma treatment, surfaces of materials are bombarded with a stream of high energy particles within the stream of plasma. Properties such as wettability, surface chemistry and surface roughness of material surfaces can be altered without the need for employing solvents or other hazardous substances. Alternative surface chemistries can be produced with plasmas, by altering the carrier gas and depositing different reactive species on the surfaces of natural fibres [56]. This can then be further exploited by grafting monomeric and/or polymeric molecules on to the reactive natural fibre surface, which can then facilitate compatibilisation with the polymer matrix.

3.3. Toughening mechanisms in PLA/wood-flour composites

Physical modification of PLA can be achieved with the incorporation of softer polymer segments, which can attach to the polymer backbone. An example of an impact modification of PLA was performed with the addition of Poly (ethylene) acrylic acid (PEAA). The effect of impact modification can be observed in the load-deflection curves depicted in Figure 4. The load-deflection curve for PLA is almost linear and displays a rapid decrease in load once the peak load is reached, which is indicative of the well-known low resistance of PLA to crack propagation and its susceptibility to brittle fracture, with the smooth impact fracture surface of PLA (Figure 5) being typical of brittle failure. The load-deflection curve for the PLA/wood-flour micro-composite containing 20%w/w wood-flour shown in Figure 4(b) displays an increased peak load compared with PLA, and the less rapid decrease in the load after peak load is reached is further evidence for effective stress transfer from the PLA matrix to the wood-flour particles. The load-deflection result for the PLA/wood-flour micro-composite containing MDI shown in Figure 4(c) indicates that the addition of MDI leads to a higher peak load compared with the equivalent micro-composite with no added MDI (Figure 4(b)), and the shape of the load-deflection curve is consistent with typical elastic-plastic deformation dominated by unstable crack growth. The increase in the peak load and width of the load-deflection profile, shown in Figure 4(d) indicates extensive plastic deformation of the PLA/wood-flour micro-composite containing PEAA. This increase in plastic deformation is attributed primarily to the increase in the rubbery nature of the blended PLA/PEAA matrix compared with PLA alone, resulting in more efficient dissipation of the energy associated with crack initiation and propagation [57].

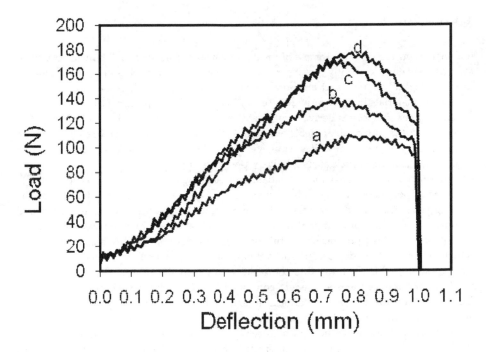

Figure 4. Load-deflection curves for a) PLA, (b) PLA/wood-flour, (c) PLA/wood-flour containing MDI, (d) PLA/wood-flour containing PEAA wood-flour content=20%w/w

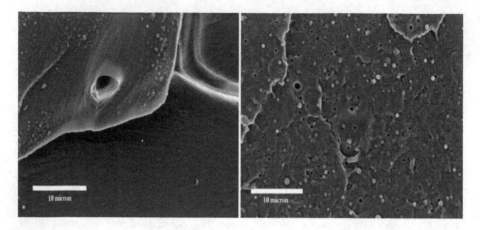

Figure 5. SEM micrograph of impact fracture surface of PLA for (a) Pure matrix and (b) blended with 3% w/w PEAA

4. Conclusions

Increasing concerns and awareness of the environment and the impact of human activity on the environment is currently the driving force for developing materials that are sustainable, more ecologically sound and are purely produced from renewable resources. PLA based composite materials have been widely recognised as a noble replacement to more conventional polymer composites derived from petrochemical feedstocks, such as polyolefins. PLA based composites are biodegradable and have high mechanical performance compared with conventional polymer composites. However, the wider uptake of PLA is restricted by performance deficiencies, such as its relatively poor impact properties which arise from its inherent brittleness, but also the limited supply and higher cost of PLA compared with commodity polymers such as polyethylene and polypropylene. It is expected that in the coming years PLA based materials will become more competitively priced as the demand increases and supply of abundant feedstock material becomes more widely available. The use of natural fibres presence a useful technique for developing PLA based composites, which are low-cost, biodegradable and can have properties that can be tailored for their specific application. PLA/natural fibre composites have been shown to have increased tensile modulus and reduced tensile strength compared with PLA, and this has been attributed to factors that include the weak interfacial interaction between the hydrophobic PLA matrix and the hydrophilic natural fibers, and lack of fiber dispersion due to a high degree of fiber agglomeration. Various methods of modifying the surface of the cellulosic fibers have been explored in an effort to improve the interaction that occurs at the interface between the PLA matrix and natural fibres. Better development of processing technologies and improvements in natural fibre treatments will facilitate the production of PLA based composites with optimum mechanical and physical performance but also generate high cost competiveness and greater acceptance of these materials in the market place.

Author details

Eustathios Petinakis[1,2*], Long Yu[1], George Simon[2] and Katherine Dean[1]

*Address all correspondence to: steven.petinakis@csiro.au

1 CSIRO, Materials Science and Engineering, Melbourne, Australia

2 Department of Materials Engineering, Monash University, Melbourne, Australia

References

[1] Mukherjee T, Kao N. PLA Based Biopolymer Reinforced with Natural Fibre: A Review. Journal of Polymers and the Environment. 2011;19(3):714-25.

[2] Psomiadou E, Arvanitoyannis I, Biliaderis CG, Ogawa H, Kawasaki N. Biodegradable films made from low density polyethylene (LDPE), wheat starch and soluble starch for food packaging applications. Part 2. Carbohydrate Polymers. 1997;33(4): 227-42.

[3] Perego G, Cella GD. Mechanical Properties. Poly(Lactic Acid): John Wiley & Sons, Inc.; 2010. p. 141-53.

[4] Bajpai PK, Singh I, Madaan J. Development and characterization of PLA-based green composites: A review. Journal of Thermoplastic Composite Materials. 2012 March 22, 2012.

[5] Yu L, Dean K, Li L. Polymer blends and composites from renewable resources. Progress in Polymer Science. [Article]. 2006;31(6):576-602.

[6] Yu T, Ren J, Li S, Yuan H, Li Y. Effect of fiber surface-treatments on the properties of poly(lactic acid)/ramie composites. Composites Part A: Applied Science and Manufacturing. [doi: 10.1016/j.compositesa.2009.12.006]. 2010;41(4):499-505.

[7] Mohanty AK, Misra M, Drzal LT. Surface Modifications of natural fibres and performance of the resulting biocomposites: An Overview. Composite Interfaces. 2001;8(5):31.

[8] Abdul Khalil HPS, Ismail H. Effect of acetylation and coupling agent treatments upon biological degradation of plant fibre reinforced polyester composites. Polymer Testing. 2000;20(1):65-75.

[9]   Sain M, Suhara P, Law S, Bouilloux A. Interface modification and mechanical proper-
      ties of natural fiber-polyolefin composite products. Journal of Reinforced Plastics and
      Composites. 2005;24(2):121-30.

[10] Huda MS, Drzal LT, Misra M, Mohanty AK. Wood-Fibre reinforced Poly(lactic acid)
      Composites: Evaluation of the Physicomechanical and Morphological Properties
      Journal of Applied Polymer Science. 2006;102:14.

[11] Xie Y, Hill CAS, Xiao Z, Militz H, Mai C. Silane coupling agents used for natural fi-
      ber/polymer composites: A review. Composites Part A: Applied Science and Manu-
      facturing. [doi: 10.1016/j.compositesa.2010.03.005]. 2010;41(7):806-19.

[12] Zafeiropoulos NE. Interface engineering of natural fibre composites for maximum
      performance. Zafeiropoulos NE, editor: Woodhead Publishing Limited; 2011.

[13] Klyosov AA. Composition of Wood-Plastic Composites: Cellulose and Lignocellulose
      Fillers. In: Klyosov AA, editor. Wood-Plastic Composites. New Jersey: John Wiley
      and Sons; 2007. p. 75-122.

[14] Mohanty AK, Misra M, Drzal LT. Sustainable Bio-Composites from Renewable Re-
      sources: Opportunities and Challenges in the Green Materials World. Journal of Pol-
      ymers and the Environment. 2002;10(1):19-26.

[15] Nzokou P, Pascal Kamdem D. X-ray photoelectron spectroscopy study of red oak-
      (Quercus rubra), black cherry- (Prunus serotina) and red pine- (Pinus resinosa) ex-
      tracted wood surfaces. Surface and Interface Analysis. 2005;37(8):689-94.

[16] Groot W, van Krieken J, Sliekersl O, de Vos S. Production and Purification of Lactic
      Acid and Lactide. Poly(Lactic Acid): John Wiley & Sons, Inc.; 2010. p. 1-18.

[17] George J, Sreekala MS, Thomas S. A review on interface modification and characteri-
      zation of natural fiber reinforced plastic composites. Polymer Engineering & Science.
      2001;41(9):1471-85.

[18] Oksman K, Skrivars M, Selin JF. Natural Fibres as reinforcement in polylactic acid
      (PLA) composites. Composites Science and Technology. 2003;63:8.

[19] Shibata M, Ozawa K, Teramoto N, Yosomiya R, Takeishi H. Biocomposites made
      from Short Abaca Fiber and Biodegradable Polyesters. Macromolecular Materials En-
      gineering. [Journal]. 2003;288:9.

[20] Wollerdorfer M, Bader H. Influence of mechanical properties of biodegradable poly-
      mers Industrial Crops and Products. 1998;8:8.

[21] Huda MS, Drzal LT, Misra M, Mohanty AK, Williams K, Mielewski DF. A Study on
      Biocomposites from Recycled Newspaper Fiber and Poly(lactic acid). Ind Eng Chem
      Res. 2005;44(15):5593-601.

[22] Bax B, Müssig J. Impact and tensile properties of PLA/Cordenka and PLA/flax com-
      posites. Composites Science and Technology. [doi: 10.1016/j.compscitech.
      2008.01.004]. 2008;68(7–8):1601-7.

[23] Mathew AP, Oksman K, Sain M. Mechanical properties of biodegradable composites from poly lactic acid (PLA) and microcrystalline cellulose (MCC). Journal of Applied Polymer Science. 2005;97(5):2014-25.

[24] Tserki V, Matzinos P, Panayiotou C. Novel Biodegradable composites based on treated lignocellulosic waste flour as filler Part II. Development of Biodegradable composites using treated and compatibilized waste flour. Composites: Part A. 2006;37:8.

[25] Petinakis E, Yu L, Edward G, Dean K, Liu H, Scully A. Effect of Matrix–Particle Interfacial Adhesion on the Mechanical Properties of Poly(lactic acid)/Wood-Flour Micro-Composites. Journal of Polymers and the Environment. 2009;17(2):83-94.

[26] Netravali AN. Fiber/Resin Interface Modification in "Green" Composites. In: Bhattacharyya SFaD, editor. Handbook of Engineering Biopolymers Homopolymers, Blends and Composites. Munich: Hanser; 2007. p. 847-68.

[27] Gandini MNBaA. The surface modification of cellulose fibers for use as reinforcing elements in composite materials. Composite Interfaces. 2004;12(1-2):41-75.

[28] Bhavesh L. Shah, Susan E. Selke, Michael B. Walters, Patricia A. Heiden. Effects of wood flour and chitosan on mechanical, chemical, and thermal properties of polylactide. Polymer Composites. 2008;29(6):655-63.

[29] Kazayawoko M, Balatinecz JJ, Matuana LM. Surface Modification and Adhesion Mechanisms in Woof Fibre-polypropylene composites. Journal of Materials Science. 1999;34:11.

[30] Lu JZ, Negulescu II, Wu Q. Maleated wood-fibre/high density polyethylene composites: Coupling mechanisms and interfacial characterization. Composite Interfaces. 2005;12(2):16.

[31] Wang Y, Yeh F-C, Lai S-M, Chan H-C, Chen H-F. Effectiveness of Functionalised Polyolefins as Compatibilizers for Polyethylene/Wood Flour Composites. Polymer Engineering and Science. 2003;43(4):13.

[32] Oksman K, Lindberg H. Influence of thermoplastic elastomers on adhesion in polyethylene-wood flour composites. Journal of Applied Polymer Science. 1998;68(11):1845-55.

[33] Sun-M. Lai F-CY, Yeh Wang, Hsun-C. Chan, Hsiao-F. Shen. Comparative Study of Maleated Polyolefins as Compatibilizers for Polyethylene/ Wood Flour Composites. Applied Polymer Science. 2003;87:487-96.

[34] X. Colom FC, P. Pages, J. Canavate. Effects of different treatments on the interface of HDPE/lignocellulosic composites. Composites Science and Technology. 2003;63:161-9.

[35] Xu T, Tang Z, Zhu J. Synthesis of polylactide-graft-glycidyl methacrylate graft copolymer and its application as a coupling agent in polylactide/bamboo flour biocomposites. Journal of Applied Polymer Science. 2012;125(S2):E622-E7.

[36] Pickering MDHBaKL. Fiber Pretreatment and Its Effects on Wood Fiber Reinforced Polypropylene Composites. Materials and Manufacturing Processes. 2006;21:303-7.

[37] Ichazo MN, Albano C, Gonzalez J, Perera R, Candal MV. Polypropylene/wood flour composites: treatments and properties. Composite Structures. 2001;54:207-14.

[38] Wu J, Yu D, Chan C-M, Kim J, Mai Y-W. Effect of fiber pretreatment condition on the interfacial strength and mechanical properties of wood fiber/PP composites. Journal of Applied Polymer Science. 2000;76(7):1000-10.

[39] Oksman MBaK. The use of silane technology in crosslinking polyethylene/wood flour composites. Composites Part A: applied science and manufacturing. 2006;37:752-65.

[40] González D, Santos V, Parajó JC. Silane-treated lignocellulosic fibers as reinforcement material in polylactic acid biocomposites. Journal of Thermoplastic Composite Materials. 2011 September 7, 2011.

[41] Hill CAS. Chemical Modification of Wood (I): Acetic Anhydride Modification 3.1. In: Hill CAS, editor. Wood Modification: Chemical, Thermal and Other Processes. New Jersey: John Wiley and Sons 2006. p. 45-76.

[42] Bogoeva-Gaceva G, Avella M, Malinconico M, Buzarovska A, Grozdanov A, Gentile G, et al. Natural fiber eco-composites. Polymer Composites. [Article]. 2007;28(1): 98-107.

[43] Li X, Tabil L, Panigrahi S. Chemical Treatments of Natural Fiber for Use in Natural Fiber-Reinforced Composites: A Review. Journal of Polymers and the Environment. 2007;15(1):25-33.

[44] Tserki V, Matzinos P, Zafeiropoulos NE, Panayiotou C. Development of Biodegradable Composites with Treated and Compatibilized Lignocellulosic Fibres. Journal of Applied Polymer Science. 2006;100:8.

[45] Doczekalska B, Bartkowiak M, Zakrzewski R. Modification of sawdust from pine and beech wood with the succinic anhydride. Holz als Roh- und Werkstoff. 2007;65(3): 187-91.

[46] Ji KLPaC. The Effect of Poly[methylene(polyphenylisocyanate)] and Maleated Polypropylene Coupling Agents on New Zealand Radiata Pine Fiber-Polypropylene Composites. Journal of Reinforced Plastics and Composites. 2004;23(18):2011-24.

[47] Bengtsson M, Oksman K. The use of silane technology in crosslinking polyethylene/ wood flour composites. Composites Part A: Applied Science and Manufacturing. 2006;37(5):752-65.

[48] Lee SH, Wang S. Biodegradable polymers/bamboo fiber biocomposite with bio based coupling agent. Composites Part A. [Article]. 2006;37(1):80-91.

[49] Wang H, Sun X, Seib P. Mechanical properties of poly(lactic acid) and wheat starch blends with methylenediphenyl diisocyanate. Journal of Applied Polymer Science. 2002;84(6):1257-62.

[50] Yu L, Dean K, Yuan Q, Chen L, Zhang X. Effect of Compatibilizer Distribution on the Blends of Starch/Biodegradable Polyesters. Journal of Applied Polymer Science. 2006;103:7.

[51] Dong S, Saphieha S, Schreiber HP. Mechanical Properties of Corona-Modified Cellulose/Polyethylene Composites. Polymer Engineering and Science. 1993;33(6):4.

[52] Dong S, Saphieha S, Schreiber HP. Rheological Properties of Corona Modified Cellulose/ Polyethylene Composites. Polymer Engineering and Science. 1992;32(22):6.

[53] Olaru N, Olaru L, Cobillac G. Plasma Modified Wood Fibres as Fillers in Polymeric Materials Romanian Journal of Physics. 2005;50(9):7.

[54] Baltazar-y-Jimenez A, Bistritz M, Schulz E, Bismarck A. Atmospheric air pressure plasma treatment of lignocellulosic fibres: Impact on mechanical properties and adhesion to cellulose acetate butyrate. Composites Science and Technology. [doi: 10.1016/j.compscitech.2007.04.028]. 2008;68(1):215-27.

[55] Lee K-Y, Delille A, Bismarck A. Greener Surface Treatments of Natural Fibres for the Production of Renewable Composite Materials Cellulose Fibers: Bio- and Nano-Polymer Composites. In: Kalia S, Kaith BS, Kaur I, editors.: Springer Berlin Heidelberg; 2011. p. 155-78.

[56] Nguyen MH, Kim BS, Ha JR, Song JI. Effect of Plasma and NaOH Treatment for Rice Husk/PP Composites. Advanced Composite Materials. [doi: 10.1163/092430411X570112]. 2011;20(5):435-42.

[57] Hristov VN, Vasileva ST, Krumova M, Lach R, Michler GH. Deformation mechanisms and mechanical properties of modified polypropylene/wood fiber composites. Polymer Composites. 2004;25(5):521-6.

Applications in Concrete Repair with FRP

Applying Post-Tensioning Technique to Improve the Performance of FRP Post-Strengthening

Mônica Regina Garcez,
Leila Cristina Meneghetti and
Luiz Carlos Pinto da Silva Filho

Additional information is available at the end of the chapter

1. Introduction

Reinforced concrete structures are, frequently, submitted to interventions aiming to restore or increase their original load capacity. According to Garden & Hollaway [1], the choice between upgrading and rebuilding is based on factors specific to each individual case, but certain issues are considered in every case. These are the length of time during which the structure will be out of service or providing a reduced service, relative costs upgrading and rebuilding in terms of labor, materials and plant, and disruption of other facilities.

Several post-strengthening techniques were developed in the last decades. Most of them are based on the addiction of a structural element to the external face of the element to be post-strengthened.

According to Täljsten [2], the method of post-strengthening existing structures with steel plates bonded to the structure with epoxy adhesive was originated in France, in the nineteen sixties, when L'Hermite (1967) and Bresson (1971) carried out tests on post-strengthened concrete beams. Additionally, Dussek (1974) reported the use of this post-strengthening method in South Africa in the middle 60's. In both cases the post-strengthening was successful and the load bearing capacity was increased. These first investigations in France and South Africa inspired future research in Switzerland (1974), Germany (1980), United Kingdom (1980), Japan (1981) and Belgium (1982). The idea of post-strengthen existing reinforced concrete structures with bonded steel was improved due to the development of synthetic adhesives, based on epoxy resins, suitable to ensure good adhesion and chemical resistance to aggressive agents.

In the last decades, non-corrosive, low-weight and high-resistant materials started to be developed and applied on the construction of new buildings, aiming to produce durable structures. These materials, called Fiber Reinforced Polymers (FRP), started to be investigated in the middle 80's at EMPA (Swiss Federal Laboratories for Materials Testing and Research), in Switzerland. At that time, the carbon fiber was elected as the most suitable for post-strengthening applications due to its low-weight, high tensile strength, high modulus of elasticity and resistance to corrosion. Since then, many structures were post-strengthened with FRP in Japan, Europe, Canada and United States and nowadays the use of FRP is growing worldwide.

Most of FRP post-strengthening systems used nowadays consist of carbon fibers embedded in epoxy matrices and provide high modulus of elasticity and tensile strength. For bridge repair, carbon fiber is the material best suited in most cases, because the fiber is alkaline-resistant and does not suffer stress corrosion, two very important arguments for such applications. Actually, there are many reasons that make carbon fibers one of the most attractive alternatives for post-strengthening concrete structures. Considering all reinforcing fiber materials used to produce FRP, the carbon fibers have the highest specific modulus and specific strength that provide a great stiffness to the system, being an ideal choice to be applied in structures sensitive to weight and deflection. Compared with steel, carbon fibers can be 5 times lighter and present a tensile strength 8 to 10 times higher.

The main impediment to the massive use of CFRP (Carbon Fiber Reinforced Polymers) regards to the high cost of the carbon fibers. Meier, in 2001 [3], pointed out that the functionality and the mechanical properties of CFRP should be better explored, due to its relatively high cost. Indeed, the use of only 10%-15% of the tensile strength of the CFRP, as it happens in some bonded post-strengthening systems, is not economically viable.

This chapter aims to analyze the efficiency of prestressed CFRP strips used to post-strengthen reinforced concrete beams, by means of cyclic and static loading tests, as an alternative to better use the tensile strength of these materials.

2. Reinforced concrete elements post-strengthened with prestressed FRP strips

The aim in prestressing concrete beams may be, according to Garden and Mays [4], either to increase the serviceability capacity of the structural system of which the beams form a part or to extend its ultimate limit state.

According to El-Hacha [5], FRP are well suited to prestressing applications because of their high strength-to-weight ratio that provides high prestressing forces, without increase on the self-weight of the post-strengthened structure. The prestressing technique may improve the serviceability of a structural element and delay the onset of cracking. When prestressed FRP are used, just a small part of the ultimate strain capacity of the material is used to prestress the FRP, the remaining strain capacity is available to support external loads and also to ensure safety against failure modes associated to peeling-off at the border of flexural cracks and at the ends of the post-strengthening.

Several FRP prestressing systems are currently available consisting of rods, strands, tendons or cables of FRP. However, in some cases, it may be advantageous to bond FRP sheets or strips onto the structural element surface in a prestressed state. According to fib Bulletin 14 [6], prestressing the FRP prior to bonding has the following advantages:

- Provides stiffer behavior as at early stages most of the concrete is in compression and therefore contributing to the moment of resistance. The neutral axis remains at a lower level in the prestressed case if compared to the unstressed one, resulting in greater structural efficiency.

- Crack formation in the shear span is delayed and the cracks, when they appear, are more finely distributed and narrower. Thus, serviceability and durability are improved, due to reduced cracking.

- The same level of strengthening is achieved with smaller areas of stressed FRP, compared to unstressed ones.

- Prestressing significantly increases the applied load at which the internal steel reinforcement begins to yield if compared to an unstressed structural member.

On the other hand, prestressing FRP systems are more expensive than the non-prestressing ones, due to the greater number of operations and the equipment that is required to prestress the FRP.

2.1. Losses of prestressing force

Prestressed FRP bonded to concrete structures are sujected to prestress losses, as it happens in any prestressing system. Such prestress losses may be instantaneous, due to immediate elastic deformation of concrete, or time dependent, due to creep and shrinkage of concrete and relaxation of the FRP.

Immediate elastic deformation of the concrete may reach 2% to 3%, according to fib Bulletin 14 [6], and happens when the prestress force is transferred into the concrete beam. If prestress is applied by reacting against the structural member there will be no loss. It happens because if the prestressing device if fixed on the structural element that will be post-strengthened, a compensation occurs: as the FRP is being stressed, the concrete is being compressed. However, FRP elements that have already been prestressed will experience a loss of prestress due to the shortening of the beam upon the prestressing of subsequent FRP elements. In such cases it is necessary to determine the average loss of prestress per FRP element.

Time dependent losses, due to creep and shrinkage of concrete, according to the fib Bulletin 14 [6], reach about 10% to 20% and are similar to the ones of conventional prestressing.

Prestressing losses due to relaxation of FRP depends, according to ACI 440.4R-04 [7], on the characteristics of the FRP composite. The document also informs that losses due to relaxation of fibers may be neglected when CFRP are used, since the relaxation of carbon fibers is very low. Losses of 0,6% to 1,2% must be considered due to the relaxation of the polymer and losses of 1% to 2% must be considered due to the straightening of fibers.

Results of a research program developed by Triantafillou et al. [8] indicate that, when pre-fabricated CFRP are used, prestress losses of 10% must be considered, due to the instantaneous and time dependent losses at the concrete and adhesive and also due to the relaxation of the CFRP.

Garden and Mays [4] consider that prestressed FRP also suffer prestress losses due to the shear transferred through the adhesive and into the concrete by the FRP tension. This shear action is sufficient to fracture the concrete even at low prestress levels so it is necessary to install anchorages at the ends of the FRP element to resist this action.

2.2. Maximum prestressing force

Figure 1(a), by Triantafillou et al. [8], shows the premature failure of a concrete beam post-strengthened with a CFRP strip, without any anchorage system, immediately after the complete release of the prestressing force. Horizontal shear cracks propagated from both ends of the CFRP strip through the concrete layer and stopped at a certain length. Figure 1 (b) shows that this failure mode may be prevented if anchorage systems are used at the ends of the strips. The authors suggest that the maximum prestressing force that avoids the need of anchorage systems provide very low prestressing levels, 15% to 20%, depending on the cross section of the CFRP strip.

(a)   (b)

Figure 1. (a) Premature failure of a prestressed CFRP strip without anchorage; (b) Action of an anchorage system (Triantafillou et al. [8]).

Thus, the addition of anchors at the end of the prestressed FRP sheets or strips reduces the shear deformation that occurs within the resin or adhesive layer upon releasing the prestressing force and reducing the shear stresses transferred to the base of the concrete section. Thereby, anchorage systems minimize the possibility of premature failures (El-Hacha [5]).

According to El-Hacha et al. [9], prestressing levels of at least 25% of the FRP tensile strength may be necessary to achieve a significant improvement in terms of the structural stiffness and load carrying capacity.

Meier [10] suggests that a prestress level as high as 50% of the CFRP strength might be necessary to increase the ultimate strength by delaying the premature failure. Experimental results presented by Deuring [11] showed that increasing the level of prestress in the CFRP from 50% to 75% reduced the strength of the beam because the highly prestressed laminates had little strain capacity remaining and the CFRP presented premature failure.

It is important to have in mind that, when post-strengthening is prestressed the modulus of elasticity of the FRP is of great significance, since the FRP element needs to be stiffer to hold up a significant loading that, before the post-strengtnening, was made only by the steel reinforcement (El-Hacha, [5]).

2.3. Prestressing techniques

Various approaches to prestress FRP have been proposed by researches and used experimentally. These methods are based on directly or indirectly prestress the FRP prior to bonding and are described bellow.

2.3.1. Cambered beam prestressing technique

In this method, developed by Ehsani & Saadatmanesh [11], no tension is directly applied to the fibers, but the FRP sheets are indirectly prestressed by cambering the beam to be post-strengthened before bonding them to the bottom face of the concrete beam.

The beam is first deflected upward by means of hydraulic jacks, as one can see in Figure 2 (a). The beam is then held in the deflected position until the adhesive is completely cured. After the cure of the adhesive, the FRP is completely bonded to the lower face of the beam and the jacks may be removed, as showed in Figure 2 (b). Once the jacks are removed, the beam will deflect downward and tensile stresses will be induced in the lower face of the beam.

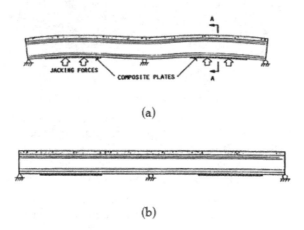

(a)

(b)

Figure 2. Sequence of prestressing procedures: (a) Camber by jacking; (b) Remove jacks when epoxy is cured (Ehsani & Saadatmanesh [11]).

The level of prestress will depend on the length of the beam and the degree of camber induced in the beam. According to Ehsani & Saadatmanesh [11], the level of prestress in this method is not high and since the highest prestress will be present at the midspan of the beam, anchorage systems are not required.

According to El-Hacha [5], in field applications, the effort required to camber the bridge near midspan is extensive relative to the low prestressing force induced in the FRP.

2.3.2. FRP prestressed against the strengthened element

In this prestressing system, developed at Queen's University and Royal Military College of Canada and presented by Wight et al. [12], the sheets are tensioned by reacting directly on the beam. The prestressed sheets are bonded to the lower face of the beam and the ends of the sheets are attached to the beam by a mechanical anchorage system. Multiple layers of prestressed sheets may be applied to the beams in successive layers, when required, due to the limits on the tensile capacity of individual sheets or due to the need to limit the load at an anchorage location.

In this process, the beams are inverted to receive the prestressing system. The mechanical prestressing and anchorage system used for the reinforced concrete beams are shown in Figure 3. According to Wight et al. [12] the mechanical anchorage system consists of steel roller anchors bonded to the sheets and steel anchor assemblies fixed to the beam. The roller anchors that grip the sheet consists of two stainless-steel rollers bonded to each end of the sheet. Prior to prestressing operations, the sheet is wrapped and bonded round the roller. To prestress the sheets, the roller at one end of the FRP sheet is fixed to the beam and the roller at the other end is movable. During prestressing, the movable roller is attached by steel prestressing strands to a hydraulic jack that reacted against the beam. The prestress is applied to the sheet, and the sliding roller is then attached, in its extended position, to a second permanent anchorage assembly. Subsequent layers may be added to the beam, using the same technique, until the desired thickness of FRP is achieved. The authors suggest that a weight perpendicular to the beam surface may be used to bring the sheet into contact with the beam surface.

El-Hacha at al. [9] used this technique to post-strengthen damaged concrete beams under severe environmental conditions. Results presented by the authors suggest that keeping the mechanical anchorage system in place prevented failures associated with high shear stresses at the ends of the sheets. Anchorages also prevented tensile fracture in the concrete cover thickness upon transfer of sheet prestress into the concrete. The prestressing system used to post-strengthen the beams improved the serviceability, controlling the formation of new cracks, delaying the formation of new cracks and limiting deflections in the beams tested. Furthermore, the prestressed CFRP sheets contributed to the load carrying capacity of the beams and significantly redistributed the stress from the internal steel reinforcement to the CFRP sheet.

2.3.3. Technique of prestressing FRP prior to bonding

In this method, studied by Triantafillou & Deskovic [13, 14], Deuring [15], Quantril & Hollaway [16] and Garden & Mays [4], the FRP sheet is first pretensioned and applied on the tensile face of the beam as one can see in Figure 4 (a) and (b). Aluminium tabs are epoxy bonded on both faces of the FRP to provide stress distribution in the end regions and then each end of the FRP is sandwiched between two predrilled steel plates bolted on at each end. The system is loaded into a prestressing frame. A vacuum bag technique may be necessary to support the external FRP during bonding.

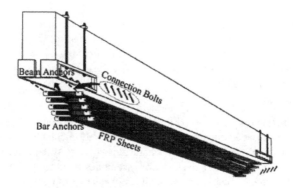

Figure 3. Prestressing system (Wight et al. [12]).

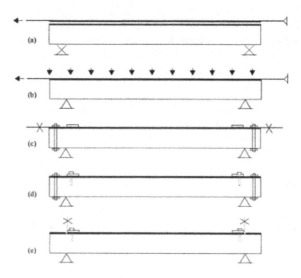

Figure 4. Sequence of prestressing procedures; (a) Pretensioning of the strip; (b) Application of the pretensioned strip on the tensile face of the beam; (c) Cutting of the sheet to transform it into a prestressing element; (d) Placement of the steel bolts; (e) End damps are removed (Garden & Mays [4]).

After the adhesive is fully cured, steel clamps are installed at each end of the beam to ensure adequate anchorage and the two ends of the FRP are cut. Then the sheet is transformed into a prestressing element as showed in Figure 4 (c).

In the next step, showed in Figure 4(d) holes are drilled through the FRP and the adhesive into the concrete beam to receive steel bolts that are bonded into the holes and allowed to cure.

Figure 4 (e) shows the last stage of the procedure, when the end clamps are removed and the FRP is cut through at each end. Then, the bolted steel endplates will play the role of the anchorage system.

Quantril & Hollaway [16] noted that cracking under the action of a given external load was found to be much less extensive and less developed than for an identical non-prestressed specimen. Besides that, prestressing produced significant increases in the load which causes yield of the internal steel over a non-prestressed specimen. Gains may be still larger if FRP action can be maintained past steel yielding to loads approaching failure. Applying the prestress prior to bonding also affects the mode of failure of the specimen, reducing the amount of shear cracking which could initiate failure in the shear spans. The greater the level of prestress, the better is the confinement effect on the development of shear cracking what increases the failure load for cases governed by shear failure. Results also suggest the levels of ductility and stiffness may be increased as well as the maximum strains in the FRP at a given load level. Despite all the advantages presented by the authors, field application of this technique would probably require methods and procedures adaptations due to working restrictions such as the overhead position and limited access of most structural elements.

According to Garden & Mays [4] the level of prestress that can be applied is limited by the tensile strength of the FRP and should not precede either yielding of the internal steel or compressive failure of the concrete to ensure adequate ductility. Results showed by the authors suggest that the level of prestress may also be limited by the strength of the plate and anchorages, by the horizontal shear strength of the adhesive-FRP interface and by the bottom layers of the concrete.

One potential benefit of this technique is the reduction of FRP material associated costs since the same strength levels can be reached with reduced area fraction (Triantafillou & Deskovic [14]). According to Triantafillou et al. [8] the method can also lend itself to prefabrication because of its simplicity and the important properties offered by FRP materials.

2.3.4. Prestressing method developed by Stoecklin & Meier [17]

Stoecklin & Meier [17] developped, at EMPA (Swiss Federal Laboratories for Material Testing and Research) a method to apply prestressed FRP strips to concrete structures. In this method, the FRP strip is first prestressed then bonded at the beam that will receive the post-strengthening. Since it is very complicated to grab and prestress the FRP strip, due to its anisotropic behavior, a prestressing device was designed, as one can see in Figure 5. The prestressing device consists of two wheels which are connected to a beam of the required length, as shown in Figure.

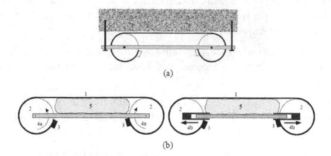

Figure 5. Prestressing device developed by Stoecklin & Meier [17]: (a) Placement of the prestressing device under the beam; (b) Two ways of prestressing a CFRP strip.

The FRP strip (1) is wrapped around the wheels (2) and clamped at its ends (3) as shown in Figure 5 (b). The strip can be prestressed by rotating one or both wheels (5a) or displace the wheels (5b). As one can see in Figure 5(a), the prestressing device with the prestressed FRP strip is temporarily mounted to the structure and can be pressed against the structure with a constant pressure by means of an air-cushion (5) between the FRP strip and the beam. (Stoecklin & Meier [17]).

In a new version of the prestressing device developed by Stoecklin & Meier [17], two separate prestressing units at each end of the strip are directly mounted to the structure, what means that the FRP strip is prestressed against the structure, as shown in Figure 6.

Figure 6. New version of the prestressing device developed by Stoecklin & Meier (Meier [10]).

To overcome anchorage problems at the ends of the FRP strips, the prestressing force can be reduced gradually from the mid-span to both ends of the FRP strips.

As described by Meier et al. [18], gradual anchoring is achieved by first bonding a fully pretensioned section in the middle of the FRP strip at mid-span. A system of electric heating may be used to speed up curing of the adhesive in the bonded section within the pot life of the adhesive. After curing the central part of the FRP strip at mid-span, the prestressing force is slightly reduced and another section is bonded at each side of the strip also using the electric heating system to speed up curing the adhesive.

This process is repeated in several stages until the entire length of the strip is bonded and the prestressed level at the ends of the strips has been reduced to a low level, as one can see in Figure 7. In this way, anchorages are not required at the end of the prestressed strip.

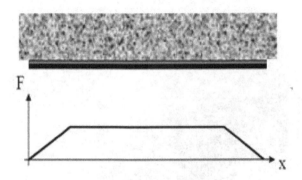

Figure 7. Gradual anchorage of prestressed CFRP strips (Stoecklin & Meier [17]).

In the prestressing method developed by Stoecklin & Meier [17] the strip is prestressed before the application at the beam. In such case, prestressing is applied by reacting against the structural member, since the prestressing device is mounted to the structure.

When the first FRP strip is prestressed, using the device developed by Stoecklin and Meier [17], imediate losses of prestress due to elastic deformation of the concrete, that happen when the prestress force is released, can be neglected, since prestressing is applied by reacting against the structural member. However, strips that have already been prestressed will experience a loss of prestress due to the shortening of the beam upon the prestressing of subsequent FRP strips.

2.4. Failure Modes of reinforced concrete beams post-strengthened with prestressed FRP submitted to static loading

According to Hollaway [19], the anisotropic behavior of the composite materials leads to a complex rupture mechanism that may be characterized by extensive damages on the composite material when submitted to static and cyclic loading. The level of damage, however, depends on the properties of the composite material and on the applied loading.

Failure modes of reinforced concrete structures post-strengthened with FRP include crushing of concrete, yielding of steel reinforcement or tensile failure at the FRP.

Teng et al. [20] report that failure modes of reinforced concrete beams post-strengthened with FRP can be broadly classified into two types: those associated with high interfacial stresses near the ends of the bonded FRP and those induced by a flexural or flexural-shear crack away from the ends, which is also referred to as intermediate crack-induced debonding. Thus premature failures, in general, are associated to:

- High interfacial stresses near to the ends of the bonded FRP, also called peeling-off.

- Flexural or flexural-shear crack away from the ends, as shown in Figure 8.

Figure 8. Rupture of FRP close to a flexural crack tip.

Concrete structures post-strengthened with prestressed FRP also show premature failures as described by Teng et al. [20]. However, in prestressed systems, the high strength of the FRP used to post-strengthen structures is much better used, and, depending on the configuration of the post.-strengthening, tensile failure of the FRP may be achieved.

Garden and Hollaway [1] presented, in 1998, a specific study regarding the failure modes of reinforced concrete beams post-strengthened with prestressed FRP, with prestressing levels ranging from 25% to 50% of the FRP strength. Results showed that a high prestress level was required to enable the ultimate capacity of strip to be reached, before shear displacement reached its critical value.

According to Garden and Mays [4], the level of prestress that can be applied will be limited by the tensile strength of the FRP. Tensile failure of the FRP should not precede either yielding of steel reinforcement or crushing of concrete, to ensure adequate ductility. Results of the experimental program showed that the level of prestress may also have to be limited by the strength of the anchorage devices, by the horizontal shear strength of the adhesive/FRP interface and by the bottom layers of concrete.

2.5. Failure of post-strengthened beams submitted to cyclic loading

Fatigue may be defined as a permanent and progressive damage process that induces gradual and cumulative crack growth and might, ultimately, result in the complete fracture of the elements subjected to cyclic loads, if the stress variation and the number of load cycles are large enough. This term was established by the first researchers of the theme due to its nature: a progressive damage process caused by cyclic loads, difficult to observe, that changes the ultimate capacity of the material (Meneghetti et al. [21]).

The usual fatigue failure mechanism for post-strengthened RC beams, when subjected to cyclic loads, is marked by the rupture of one of the steel rebars, followed by a stress redistribution that overloads the remaining bars.

Meier U. [22] highlights that the steel rebars fail before the FRP post-strengthening, as can be seen in Figure 9 that shows the steel rebars of a post-strengthened concrete beam after a fatigue failure. However, cyclic loads can also damage the adhesive and affect the interface concrete-adhesive and adhesive-FRP, leading to premature failures.

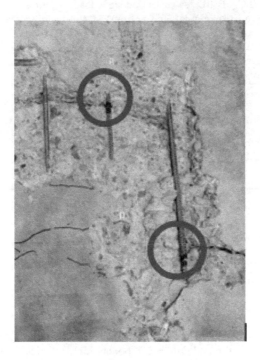

Figure 9. Failure of concrete beam after fatigue loading test.

According to Ferrier et al. [23], the performance and the durability of a concrete structure post-strengthened with FRP, when subjected to cyclic loads, depends not only on the FRP fatigue behavior but also depends on the interface concrete-adhesive and adhesive-FRP. Authors point out the importance of understanding the behavior of these materials under cyclic loading, since a typical reinforced concrete highway bridge deck with a design life of 40 years may experience a minimum of 58×10^8 loading cycles of varying intensities.

3. Experimental analysis of reinforced concrete beams post-strengthened with prestressed FRP

3.1. Description of specimens

Aiming to analyze the behavior of concrete beams post-strengthened with prestressed CFRP strips under static loading three beams were tested: VT, VFC_NP_01 VFC_PE_01 (Table 1). Regarding the cyclic loading tests two beams were tested: VFC_PC_01 and VFC_PC_02 (Table 2). Stress levels applied at beam VFC_PC_01 were 50% and 80% of the yielding stress observed at beam VFC_PE_01, tested under static loading. Stress levels applied at beam VFC_PC_02 were more reasonable, 50% and 60% of the yielding stress observed at beam VFC_PE_01.

Beam	Post-strengthening	Prestressing level
VT	-	-
VFC_NP_01	Two CFRP non-prestressed strips	-
VFC_PE_01	Two CFRP prestressed strips	35% of ε_{fu}

Table 1. Description of experimental program – static loading.

Beam	Post-strengthening	Test	Prestressing level applied on the strips	Stress range of fatigue loading
VFC_PC_01	Two CFRP prestressed strips	Bending cyclic loading	35% of ε_{fu}	50% to 80%
VFC_PC_02				50% to 60%

Table 2. Description of experimental program – cyclic loading.

3.2. Reinforced concrete beams

The reinforced concrete beams were rectangular, 6500mm long, 1000mm wide, and 220mm deep. All beams were reinforced with seven bottom 15mm steel bars ($\varrho = 0.0041$). The shear reinforcement consisted of 8mm steel stirrups spaced each 90mm ($11.17 \text{cm}^2/\text{m}$). Geometry and reinforcement details for the beam are shown in Figure 10.

Aggregates used to produce the concrete were the ones available in the Switzerland region and the cement was the Portland CEM I 42.5 (95% of clinquer, and 5% of other components), equivalent to the Brazilian CPI. The average compressive stress (cube strength) of the concrete, after 28 days, was 44MPa.

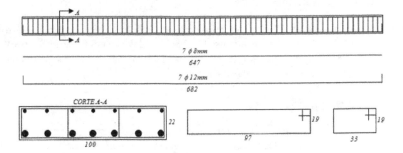

Figure 10. Geometry and reinforcement details of tested beams.

	Carbodur S 512 Strip	
Thickness (mm)	1.2	
Width (mm)	50	
Tensile Strength (MPa)	2,800	
Ultimate Strain ($^o/_{oo}$)	17	
Young's Modulus (MPa)	165,000	
Temperature Resistance (°C)	150	
Fiber Volumetric Content (%)	68	
Density (g/cm³)	1.60	
	Sikadur®-30 Resin	Sikadur®-30LP Resin
Components	3Part A:1Part B	2Part A:1Part B
Pot life at 25°C (min)	-	60
Pot life at 35°C (min)	40	-
Pot life at 55°C (min)	-	30
Tg (°C)	62	107
Young's Modulus (MPa)	12,800	10,000

Table 3. Post-strengthening system characteristics.

The 8mm steel bars had average yield stress, yield strain, ultimate stress and modulus of elasticity of 554MPa, 2.51°/oo, 662MPa and 220 Gpa, as indicated by tensile tests. The 12mm steel bars had average values of: 436MPa, 1.98°/oo, 688MPa e 215 GPa.

3.3. Post-strengthening system

Sika® Carbodur (Carbodur S 512 and Sikadur®-30) was the CFRP system used to post-strengthen the beams. However, Sikadur®-30LP adhesive was used to bond the prestressed strips to the concrete, due to its extended pot life. Table 3 shows the characteristics of the strips and adhesives, provided by manufacturer.

3.4. Post-strengthening procedure

The application of Sika® Carbodur system demands a surface preparation for the concrete and the strip. The concrete surface must be clean and free from grasses and oil, dry and have no loose particles. Considering the application of prestressed strips, after concrete and strip surface preparation, the strip is clamped, prestressed and covered with Sikadur®-30LP adhesive. Then, thermocouples are settled at the strip aiming to control the temperature applied to accelerate the cure of the adhesive. Figure 11 shows the application of epoxy adhesive on the strip and the procedure to clamp the strip on the prestressing device.

A gradual anchorage system was applied: after the cure of the adhesive at the middle part of the beam, the prestressing force was marginally reduced and the following areas were bonded. The force was reduced further and the adjacent areas were bonded. This procedure was repeated until there was no remaining prestressing force at the ends of the strip. With the gradual reduction, the level of prestress applied at the ends of the strip is very low or close to zero, eliminating the need of additional anchorage systems.

The maximum prestressing force applied to prestress the strip was 60kN, at mid-span. Then, prestressing force was gradually reduced, to 48kN, 36kN, 24kN, 12kN and, finally, zero, at the ends.

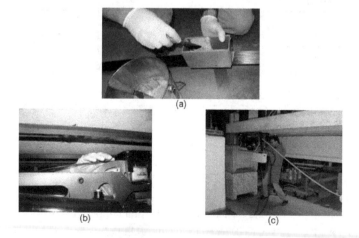

Figure 11. (a) Application of epoxy adhesive; (b) e (c) Procedure to clamp the strip on the prestressing device.

3.5. Test Procedure

The experimental program was developed at EMPA – Switzerland. Loading was applied according to a six point bending test scheme: simple supported beam and four vertical loads, spaced by 12000mm, symmetrically applied along the 6000mm span. Load was applied by two 100kN hydraulic jacks. During the tests, values of deflection at mid-span and specific strain at steel, concrete and FRP were continuously recorded by a computer controlled data acquisition system.

3.6. Behavior of post-strengthened beams tested under static loading

3.6.1. Loads and failure modes

Table 4 shows that the flexural capacity of beam VFC_NP_01, post-strengthened with two non-prestressed CFRP strips, increased 27% when compared to the control beam. On the other hand, post-strengthening of beam VFC_PE_01, two prestressed CFRP strips, increased 62.41% the load bearing capacity of the beam.

Beam	Post-strengthening	Ultimate load	Failure Mode
VT	-	100.14kN	Yielding of steel followed by concrete crushing
VFC_NP_01	Two 1.2mm x 50mm non-prestressed strips	127.25kN	Premature failure (peeling-off)
VFC_PE_01	Two 1.2mm x 50mm prestressed strips	162.41kN	Premature failure (peeling-off)

Table 4. Ultimate Loads and failure modes of post-strengthened beams tested under static loading.

(a)   (b)

Figure 12. Beam VFC_NP_01: (a) During test; (b) After premature failure of strips.

Results of beams VFC_NP_01 and VFC_PE_01 can be explained by the principles of pre-stressing. When the prestressing force applied on the CFRP strips is released, compressive stresses are induced on the concrete. Such compressive stresses delay the concrete cracking and the yielding of the steel reinforcement. Thus, the load bearing capacity of the post-strengthened element is increased.

Premature failures (peeling-off) of beams VFC_NP_01 and VFC_PE_01 occurred due to the high interfacial stresses near to the ends of strips. Peeling-off failures are catastrophic and happen without any previous advice. Figure 12 shows two CFRP strips of beam VFC_NP_01 after peeling-off. Both strips are completely detached from the beam; however, strips do not present any damage, once the failure occurred at the concrete/adhesive interface.

3.6.2. Displacements at mid-span

Figure 13 shows that all post-strengthened beams present similar behavior regarding stiffness until concrete cracking. Results indicate that in such cases the action of the post-strengthening begins just when the structural element is already cracked.

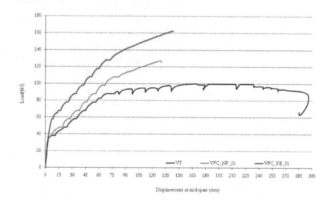

Figure 13. Load vs Displacement response of the beams tested under static loading.

Beam VFC_NP_01 post-strengthened with non-prestressed strips and the control beam, showed no significant difference regarding stiffness. However, beam VFC_PE_01 showed a stiffer behavior, when compared to beans control and VFC_NP_01, due to the increase of the cracking load and the later yielding of the reinforcement steel.

Control beam and beam VFC_NP_01, post-strengthened with non-prestressed CFRP strips present similar values of loading and displacement at mid-span at concrete cracking, due to the fact that the post-strengthening begins to act just after concrete cracking.

Cracking load of beam VFC_PE_01 was 57% higher than the ones of control beam and beamVFC_NP_01. Differences at the first stage of the loading versus displacement response at mid-span happen because prestressing leads to the development of compressive stresses at the bottom face of the beam. When non-prestressed strips are applied, the bottom face of the beam is already tensioned at the beginning of the loading process. Thus, when post-strengthening is prestressed, the loading needed to crack the concrete is significantly higher.

Differences regarding yielding of the reinforcement steel were significant. Increasing of 22% for beam VFC_NP_01 and 45% for beam VFC_PE_01, related to control beam, was observed.

3.6.3. Anchorage system

The gradual anchorage system worked properly and allowed the use of 83% of the tensile strength of the strips. Advantages of the gradual anchorage include the fact that the same device is used to prestress the strip and to gradually reduce the prestress load from the mid-span to the ends of the strip. The result is a prestressed CFRP strip without any external anchorage systems.

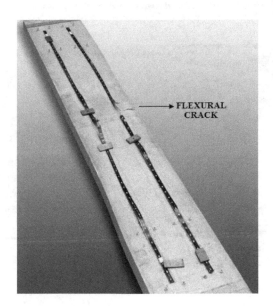

Figure 14. Beam VFC_PC_01 after failure.

3.7. Behavior of post-strengthened beams tested under cyclic loading

3.7.1. Beam VFC_PC_01

Beam VFC_PC_01 was submitted to stress levels of 50% and 80% of the yielding stress observed at beam VFC_PE_01, tested under static loading. Maximum and minimum applied loads were, respectively, 80kN and 40kN, which, added to the self weight, 28 kN, resulted applied loads of 108kN and 68kN (66% and 42% of the ultimate capacity of beam VFC_PE_01). Aiming to produce the first cracks, beam VFC_PC_01 was first pre-loaded up to108kN. Then, cyclic loading was applied at a frequency of 4Hz.

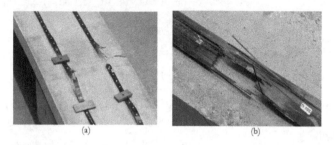

Figure 15. Failure of post-strengthening strips: (a) Next to a flexural crack; (b) Distant from flexural cracks.

Figure 16. Fatigue failure of steel rebars.

When 282,000 cycles were reached, a crack of about 2.2mm was observed, approximately at mid-span, reaching about 90% of the cross-section. After 331,300 cycles the machine automatically stopped, when the deflection limit was reached. It was not observed any sign of apparent failure at the strips. Larger displacement limits were settled and the test was restarted. However, the post-strengthening failed before the maximum load of 108kN was reached. When the first strip debonded, the sudden release of the pre-tensioning force caused a compressive failure at the CFRP, as a secondary failure (Figure 14). The secondary failure occurred in a region that was damaged due to the presence of the flexural crack showed in Figure 14, which reached about 90% of the cross-section.

Figure 15 shows one of the CFRP strips after the failure of the post-strengthened beam. Flexural crack showed in Figure 14 induced the identification of a steel rebar broken due to fatigue. After the test, the concrete was removed from the bottom of the beam and all steel rebars were inspected. Figure 16 confirms the existence of more steel rebars broken also due to fatigue.

Figures 17 and 18 show strains in the concrete and in the CFRP strips, obtained by deformeters, which gauge points were placed along the bottom of the beam. Measurements were made during pre-loading, after 30,000 cycles, and after 100,000 cycles, with the beam subjected to the maximum load (108kN). It is shown that the measurements made in the CFRP strips allowed the construction of well defined curves. However, due to the crack growing between the gauge points, this behavior could not be observed in the measurements made in the concrete. Nevertheless all obtained responses followed an expected pattern.

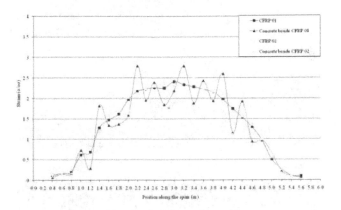

Figure 17. CFRP and concrete strains of beam VFC_PC_01, submitted to 108kN, during pre-loading.

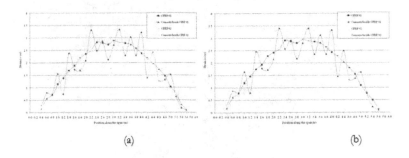

(a)   (b)

Figure 18. CFRP and concrete strains of beam VFC_PC_01, submitted to 108kN after (a) 30,000 cycles and (b) 100,000 cycles.

Strains in the CFRP strips, at mid-span, during pre-loading, varied from 2.00º/oo up to 2.50º/oo. At the end of 30,000 cycles, strains increased to levels that varied from 2.50º/oo up to 3.00º/oo. From 30,000 cycles up to 100,000 cycles, it was not observed any significant variation in the strains. Strains measured, added to the strain applied to prestress each strip (5.95º/oo), give for each strip a total strain of of 8.45º/oo and 8.95º/oo. It is also noted that, at mid-span, where most of the cracks could be found, strains measured in the concrete and in the FRP are quite different. It happens because the FRP strip acts as a belt, blocking the concrete crack opening. Therefore, it can be observed that several points along to the beam are subjected to different strains. Such points can, eventually, be related to the occurrence of premature failures. Strains at a distance of 1,2 m from both beam ends, out of the loading region, are not greater than 1.50º/oo, however, these values increased about 100% from pre-loading up to 100,000 cycles.

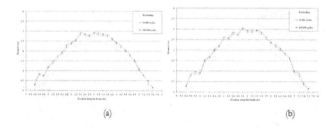

(a)   (b)

Figure 19. CFRP strains of strip 01 (a) and 02 (b) of beam VFC_PC_01 during cyclic loading.

Figure 20. Cracks at mid-span of beam VFC_PC_01 after 100,000 cycles.

Figure 19 shows the strains in the CFRP strips measured during pre-loading, after 30,000 cycles and after 100,000 cycles. Significant variations were not observed in the range from 30,000 to 100,000 cycles. However, strains increased about 0.50º/oo from the pre-loading up to 10,000 cycles. Figure 19 also shows that up to 100,000 cycles the behavior of both strips is similar. However, some variations can be observed at mid-span, due to the high cracking.

The strategy adopted to monitor the crack growing at mid-span of beam VFC_PC_01 can be observed at Figure 20. Results of crack openings and the respective position from the left end of the beam are shown in Figure 21.

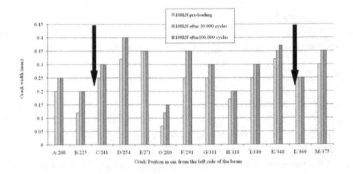

Figure 21. Cracks at mid-span of beam VFC_PC_01 after 100,000 cycles.

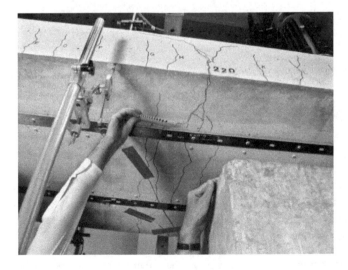

Figure 22. 2,2mm width crack at mid-span of beam VFC_PC_01 after post-strengthening failure.

A high concentration of cracks can be observed at the mid-span of beam VFC_PC_01, be-tween the four loading points. Figure 21 shows the crack openings at mid-span between the two central loading points (signaled on the figure by two vertical arrows). It is noteworthy that, before 100,000 cycles, crack openings did not reached 0.05mm.

The highest crack opening after 100,000 cycles, named D, was 0.4mm, located 254cm from the left side of the beam. However, apparently the post-strengthening failed due to a 2.2mm crack opening, named I, after 331,300 cycles. It is important to notice that, after 100,000 cy-cles, this crack opening was about 0.3mm (Figure 22).

The decision of testing beam VFC_PC_01 under a high stress variation led to the fatigue fail-ure before 5,000,000 cycles, that was considered the pattern of infinite fatigue life. Stress lev-els applied to the beam VFC_PC_02, however, are more consistent with the ones usually found in real structures. Results of beamVFC_PC_02 will allow a more detailed analysis of the CFRP prestressing technique used, as well as of the gradual anchorage system.

3.7.2. Beam VFC_PC_02

Beam VFC_PC_02 was submitted to stress levels of 50% and 60% of the yielding stress ob-served at beam VFC_PE_01, tested under static loading. Maximum and minimum applied loads were, respectively, 50kN and 40kN, which, added to the self weight of 28 kN, resulted in applied loads of 78kN and 68kN (48% and 42% of the ultimate capacity of VFC_PE_01). Aiming to produce the first cracks, beam VFC_PC_02 was pre-loaded up to78kN. Then, cy-clic loading was applied at a frequency of 4Hz.

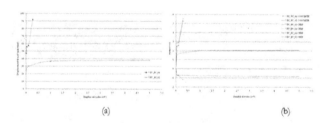

(a)   (b)

Figure 23. Beams VFC_PC_01 and VFC_PC_02: (a) Displacement at mid-span vs number of cycles (b) Concrete and CFRP strains vs number of cycles.

Figure 23 (a) shows that vertical displacements at mid-span, for beam VFC_PC_02, meas-ured with the beam subjected to the maximum load (78kN) varied 12.70mm from preload-ing up to 1,000,000 cycles. From this point, up to the end of the test, vertical displacements increased just 1.49mm. Data of beam VFC_PC_01 showed a large increase in the vertical dis-placements at mid-span after 100,000 cycles, probably due to the fatigue failure of the steel rebars. Strains in the concrete and in the CFRP strips (Figure 23 (b)) behave similarly to the displacements at mid-span, where most of the variations occurred before 1,000,000 cycles, and, after that, showed stability up to 5,000,000 cycles. Beam VFV_PC_01 also showed a similar behavior between strains and vertical displacements at mid-span, however, with a significant increasing after 100,000 cycles, probably due to the fatigue failure of the steel rebars.

Figures 24 to 26 show the strains in concrete and in CFRP strips, obtained by deformeters, which gauge points were placed along the bottom of the beam. Measurements were made during pre-loading, after 30,000 cycles, after 100,000 cycles, after 1,000,000 cycles and after 5,000,000 cycles, with the beam subjected to the maximum load (78kN).

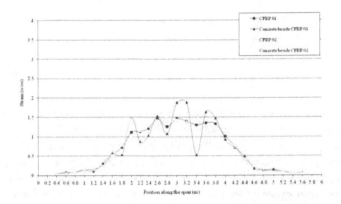

Figure 24. CFRP and concrete strains of beam VFC_PC_02, submitted to 78kN, during pre-loading.

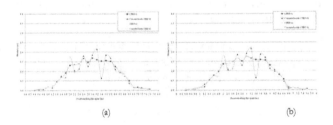

Figure 25. CFRP and concrete strains of beam VFC_PC_02, submitted to 78kN after (a) 30,000 cycles and (b) 100,000 cycles.

It can be noticed that the strains obtained by deformeters, placed along the bottom of the beam VFC_PC_02, varied from the pre-loading up to 1,000,000, tending to stabilize after 5,000,000, cycles. Portions located between the loading points (1,2 m to 4,8 m from the beam end) clearly show the presence of cracks in the concrete. From the pre-loading up to 30,000 cycles, it was not observed any significant variation in the strains along the gradual anchorage zone (1.2m from the both beam ends). Strains increased after 100,000 cycles, and, after 5,000,000 cycles, the level of strains of the beginning of the test could be observed just along the first 0.6m from both beam ends. Figures 24 to 26 show that strains in the anchorage zones of beam VFC_PC_01 were higher than the ones of beam VFC_PC_02.

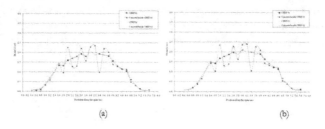

(a)   (b)

Figure 26. CFRP and concrete strains of beam VFC_PC_02, submitted to 78kN after (a) 1.000.000 cycles and (b) 5.000.000 cycles.

Data of Figures 24 to 26 show that strains in the CFRP strips, between the two central loading points, placed at 2.4m and 3.6m from the beam ends, measured at pre-loading, varied from 1.00 º/oo up to 1.50º/oo. After 30,000 cycles strains increased up to 2.00 º/oo and no significant variation were observed from 30,000 cycles up to 100,000 cycles. Measurements after 1,000,000 and 5,000,000 cycles registered a maximum strain of 2.11º/oo. Such strain, added to the strain applied to prestress each strip (5.95º/oo), give for each strip a total strain of of 8.06º/oo. Strains in the strips of beam VFC_PC_02 were smaller to the ones of beam VFC_PC_02, since, for the second beam, the maximum load and the difference between the maximum and the minimum load were smaller. Results of beam VFC_PC_02 indicate that, up to 5,000,00 cycles, it was not observed any damage on the post-strengthening system, due to the application of the cyclic loading.

Figure 27 (a) shows the strains in the CFRP strips, measured from the pre-loading up to 5,000,000 cycles. The most significant variations occurred up to 1,000,000 cycles. Strains in the CFRP strips varied about 0.85º/oo from the pre-loading up to 5,000,000 cycles. The greatest differences regarding strains were found at 1.8m, 2.8m and 4.6m from the left side of the beam (Figure 27 (a)), and at 2.6m from the left side of the beam (Figure 27 (b)).

Results indicate the existence of a kind of progressive strain at the anchorage regions, which can, ocassionally, generate adherence problems regarding long-term fatigue. Such effect should be better investigated, however, the long time demanded to realize fatigue tests, sometimes, inhibits this initiative.

Crack growing at mid-span of beam VFC_PC_01 can be observed at Figure 28, which shows the results of all crack opening measurements made, from pre-loading up to 5,000,000 cycles. Figure 28 shows the results of crack openings and the respective position from the left side of the beam, at mid-span, between the two central loading points (signaled on the figure by two vertical arrows).

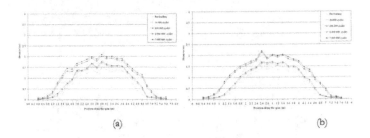

(a)   (b)

Figure 27. CFRP strains of strip 01 (a) and 02 (b) of beam VFC_PC_02 during cyclic loading.

First cracks at mid-span appeared during pre-loading, reaching less than 0.15mm. From this point up to 100,000 cycle, crack openings increased, but did not exceed 0.20mm. After 1,000,000 cycles the maximum crack opening was 0.22mm, and after 5,000,000, this value was not exceeded. Cracking at the gradual anchorage regions appeared just after 100,000 cycles, however, the maximum cracking opening observed was 0.05mm. From 100,000 up to 5,000,000 cycles, the maximum crack opening measured at these regions was 0.10mm. Results of crack openings obtained from beams VFC_PC_01 and VFC_PC_02 cannot be compared directly, due to the difference regarding the maximum and minimum loads applied to generate the cyclic loading. As the maximum load applied on the beam VFC_PC_01 (108kN) was higher than the one applied on the beam VFC_PC_02 (78kN), beam VFC_PC_01 showed higher values of crack openings once pre-loading. Values of crack openings obtained after 5,000,000 cycles, for beam VFC_PC_02, were reached by beam VFC_PC_01 after just 282,000 cycles.

4. Conclusions

4.1. Post-strengthened beams tested under static loading

Results obtained with the development of the research program allowed the investigation of changes on the behavior of post-strengthened elements due to prestressing. The increasing on the load bearing capacity of the beam post-strengthened with prestressed strips, higher than the one of the beam post-strengthened with non-prestressed strips, highlights the efficiency of the prestressing technique. All post-strengthened beams showed vertical displacements at mid-span lower then the ones of the control beam. However, the stiffer behavior showed by all post-strengthened beams was evidenced only after concrete cracking. Due to the increasing of the concrete cracking load and the later yielding of the reinforcement steel, the beam post-strengthened with prestressed CFRP strips showed a stiffer behavior when compared to the one post-strengthened with non-prestressed CFRP strips. Gradual anchorage worked properly, dismissed the use of any external anchorage system and allowed the use of 83% of the tensile strength of the strip.

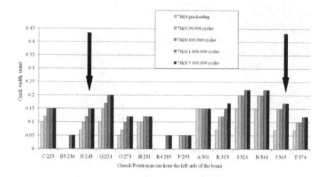

Figure 28. Cracking at mid-span of beam VFC_PC_2 during test.

4.2. Post-strengthened beams tested under cyclic loading

Results of the reinforced concrete beams tested under cyclic loading show that when these structures are post-strengthened with prestressed CFRP strips, damages that occur due to fatigue are mainly related to the level of stress at the steel rebars. Experimental results showed that the damage, which led to the rupture of the steel rebars, is related to the level of stress during loading, and, that it is not related to the type of post-strengthening. Tests showed that an increasing of 20% in the maximum stress of the steel rebars significantly reduced the fatigue life time of the post-strengthened element, decreasing about 15 times the number of cycles up to failure. These results emphasize the importance of proceeding the monitoring of structures that are usually submitted to cyclic loading, such as highway and railway bridges. In some cases, when these structures were designed to support traffic loads smaller to the ones that they are submitted nowadays, the use of post-strengthening may increase their lifetime, since the use of post-strengthening may lead to a reduction in the stress level of the steel rebars.

Acknowledgements

Authors would like to acknowledge CNPq (Portuguese acronym of the Brazilian Ministry of Science's National Research Council) and CAPES (Portuguese acronym of the Higher Education Human Resources Development Agency of the Brazilian Ministry of Education) for providing the financial support needed to develop this project. Authors would also like to express their appreciation for the technical support given by the research team EMPA (Swiss Federal Laboratories for Materials Testing and Research), in Switzerland. Finally, authors gratefully acknowledge Prof. Urs Meier for his support and assistance during the PhD. studies of the first author of this paper, developed at EMPA – Switzerland.

Author details

Mônica Regina Garcez[1*], Leila Cristina Meneghetti[2] and Luiz Carlos Pinto da Silva Filho[3]

*Address all correspondence to: mrgarcez@hotmail.com

1 Federal University of Pelotas, Brazil

2 University of São Paulo, Brazil

3 Federal University of Rio Grande do Sul, Brazil

References

[1] Garden, H. N., & Hollaway, L. C. (1998). An experimental study of the failure modes of reinforced concrete beams strengthened with prestressed carbon composite plates. *Composites Part B*, 29(B), 411-424.

[2] Täljsten, B. (1994). Strengthening of existing concrete structures with epoxy bonded plates of steel or fibre reinforced plastics. *PhD thesis. Lulea University.*

[3] Meier, U. (2001). Polyfunctional use of advanced composite materials with concrete. *N. Banthia, N, Sakai K, Gjørv OE. (eds.) CONSEC01: proceedings of the Third International Conference on Concrete Under Severe Conditions, CONSEC01, Vancouver, Canada. Vancouver: Vancouver B. C.*

[4] Garden, H. N., & Mays, G. C. (1999). Strengthening of Reinforced Concrete Structures Using Externally bonded FRP Composites in Structural and Civil Structures. *Cambridge: CRC Press.*

[5] El-Hacha RMA. (2000). Prestressed CFRP for strengthening concrete beams at room and low temperatures. *PhD thesis, Queen's University .*

[6] Fédération Internationale du Betón. (2001). Externally bonded FRP reinforcement for RC structures- Bulletin 14. *Lausane: FIB;. Technical Report.*

[7] American Concrete Institute. (2004). Prestressing Concrete Structures with FRP Tendons. *Farmington Hills: ACI,*, 35, ACI 440.4R-04.

[8] Triantafillou, T. C., Deskovic, N., & Deuring, M. (1992). Strengthening of concrete structures with prestressed fiber reinforced plastic sheets. *ACI structural Journal:*, 89(3), 235-244.

[9] El -Hacha, R., Wight, G., & Green, M. F. (2004). Prestressed carbon fiber reinforced polymer sheets for strengthening concrete beams at room and low temperatures. *Journal of Composites for Construction:*, 3-13.

[10] Meier, U. (2005). Design and rehabilitation of concrete structures using advanced composite materials. *Klein GMB. (ed.) PRECONPAT05: proceedings of the Pre-Congresso Latino-Americano de Patologia da Construção, PRECONPAT05, Porto Alegre, Brazil. Porto Alegre: UFRGS.*

[11] Ehsani, M. R., & Saadatmanesh, H. (1989). Behaviour of externally prestressed concrete girders. *Iffland JSB. (ed.) STRUCTURES CONGRESS 89: proceedings of the Seventh Annual Structures Congress, STRUCTURES CONGRESS 89, San Francisco, United States. New York: ASCE.*

[12] Wight, R. G., Green, M. F., & Erki, MA. (2001). Prestressed FRP sheets for post-strengthening of reinforced concrete beams. *Journal of Composites for Construction:,* 214-220.

[13] Triantafillou, T. C., & Deskovic, N. (1991). Innovative prestressing with FRP sheets: Mechanics of short-term behavior. *Journal of Engineering Mechanics:,* 117(7), 1652.

[14] Triantafillou, T. C., & Deskovic, N. (1992). Presressed FRP sheets as external reinforcement of wood members. *Journal of Structural Engineering:,* 118(5), 1270.

[15] Deuring, M. (1993). Verstärken von Stahlbeton mit gestpannten Faserverbund-Werkstoffen. *Dübendorf: EMPA, 276, Report 224.*

[16] Quantryll, R. J., & Hollaway, L. C. (1998). The flexural rehabilitation of reinforced concrete beams by the use of prestressed advanced concrete plates. *Composites Science and Technology,* 58, 1259.

[17] Stöcklin, I., & Meier, U. (2003). Strengthening of concrete structures with prestressed and gradually anchored CFRP. *Tan KH. (ed.) FRPRCS-6: proceedings of the Sixth International Symposium on FRP Reinforcement for Concrete Structures, FRPRCS-6, Singapore. Singapore:World Scientific.*

[18] Meier, U., Stöcklin, I., & Terrasi, G. P. (2001). Making better use of the strength of advanced materials in structural engineering. *Teng JG. (ed.) CICE01: proceedings od the FRP Composites in Civil Engineering, CICE01, Hong Kong, China. Hong Kong: Elsevier.*

[19] Hollaway, L. (1993). Polymer Composites for Civil and Structural Engineering. *Hong Kong: Blackie Academic and Professional.*

[20] Teng, J. G., Smith, S. T., Yao, J., & Chen, J. F. (2003). Intermediate crack-induced in RC beams and slabs. *Construction and Building Materials:,* 17-447.

[21] Meneghetti, L. C., Garcez, M. R., Silva, Filho. L. C. P., & Gastal, F. P. S. L. (2011). Fatigue life regression model of reinforced concrete beams strengthened with FRP. *Magazine of Concrete Research:,* 63-539.

[22] Meier, U. (1995). Strengthening of structures using carbon fibre/epoxy composites. *Construction and Building Materials:,* 9(6), 341-351.

[23] Ferrier, E., Bigaud, D., Hamelin, P., Bizindavyi, L., & Neale, K. W. (2005). Fatigue of CFRPs externally bonded to concrete. *Materials and Structures:,* 38(1), 39-46.

Chapter 4

The Use of Fiber Reinforced Plastic for The Repair and Strengthening of Existing Reinforced Concrete Structural Elements Damaged by Earthquakes

George C. Manos and Kostas V. Katakalos

Additional information is available at the end of the chapter

1. Introduction

During the last fifty years various parts of the world have been subjected to a number of damaging earthquakes. Greece is one of the countries where such damaging earthquakes occur quite frequently ([1]). Some of these earthquakes, not necessarily the most intense, occurred near urban areas and thus subjected various types of structures to significant earthquake forces leading to damage ([2]). For some of these earthquakes, ground motion acceleration recordings were obtained at distances relatively close to the area of intense shaking, thus providing valuable information for correlating the observed damage with this ground motion recording and its characteristics. Moreover, following the most damaging of these earthquakes, studies were initiated that led to the revision of the provisions of Seismic Codes [3]. The damaged structures included old structural formations, sometimes older than one hundred years, which were not designed for seismic forces. They usually belong to cultural heritage and are under various forms of conservation status that does not allow all types of retrofitting but only retrofitting materials and techniques that are compatible with the existing materials; moreover, the applied retrofitting in these cases must also be reversible so that it can be easily removed in case it demonstrates undesirable effects with time. Apart from the cultural heritage structures, the damaged structures also include relatively contemporary structures that are usually less than fifty years old. The vast majority of these structures are multistory reinforced concrete (R/C) buildings. There are other types of structures apart from (R/C) buildings, such as structures forming the infrastructure or industrial facilities which can also develop earthquake damage. However, this chapter will be devoted to the usual R/C residential multi-story buildings, the earthquake damage of their structural elements and their strengthening.

1.1. Damage observations for contemporary R/C structures

These structures are usually designed and built according to the provisions of a Seismic Code ([3]). The cause of damage may be due to:

The code provisions underestimating the severity of the shaking and thus underestimating the seismic demand posed upon the various structural elements, their connections and the foundation of the whole structure.

The code provisions together with the specification of the materials resulting in such strengths that are below the seismic demands placed upon the structural elements.

The detailing and the realization of the design during construction or alteration during the life time of the building resulting again in such strengths for the various structural elements that are below the corresponding seismic demands placed upon them.

In all cases the appearance of structural damage results from the above fundamental causes, either from one of them or from their combination; this is expressed by the following inequality (Eq. 1) between the strength and the demand put upon the various members of the structural system, whereby the demand is signified as S_d and the strength as R_d. Damage is expected to occur when this inequality is not satisfied [4].

$$S_d < R_d \qquad\qquad (1)$$

A serious task, after a strong earthquake affects an urban area, is to appraise the severity of the structural damage for a large number of buildings. This is usually done in Greece by engineers specially appointed for this task by the Government. The screening process commences a few days after the natural disaster, provided that the seismic sequence has subsided. There are certain guidelines published by the Hellenic Organization of Earthquake Planning and Protection to facilitate the appointed engineers in this damage screening process [5]. In general terms the first round of inspection must lead to the classification of each building to one of the three main categories. The first damage category is that there is no structural damage and the mainly non-structural damage does not pose any danger; so these buildings can be reused immediately. The second damage category is that there is non-structural damage as well as some structural damage; the latter, although contained, may have led to a considerable decrease of the bearing seismic capacity of the damaged structural elements and the structure as a whole. There may be need for temporary shoring and the removal of dangerously damaged non-structural elements. These buildings will be subjected to a second round of inspection after these countermeasures have been accomplished. Their reuse will be decided after the second round of screening. The third category includes buildings with relatively extensive damage to its structural elements (slabs, beams, columns or shear walls) [5], [6]; the damage is relatively widespread in terms of story level. Permanent deformations of the structural elements are evident in the form of concrete cracking and crushing in certain critical areas of the structural elements indicating that these areas have been overstressed and inequality 1 has ceased to be valid. There is serious consideration that this bearing seismic capacity reduction may lead to partial collapse. There may be urgent need for temporary shoring and the removal of dangerously damaged parts in order to avoid, if possible, partial collapse (figure 1b). These buildings will also be subjected to a second round of inspection. If it is decided that they pose a public threat, due to the possibility of partial or total collapse, they should be demolished (figure 1a); otherwise, a special design should devise a feasible scheme for their repair and strengthening. The following sections will address the task of repair and strengthening of structural elements that have been subjected to such earthquake damage to their structural system and have been classified as belonging to the second or third seismic damage categories after the second round of screening process.

a   b

Figure 1. a The old building of a pharmaceutical company that collapsed during the Athens 1999 earthquake together with an upgraded building next to it that survived unscathed [2]. b. Temporary shoring of 4-story building (Hardas) damaged during the Pyrgos 1993 earthquake [2].

1.2. Structural damage description at the level of the structural element

It is usual to describe the structural damage at the level of each reinforced concrete structural member, e.g. slab, beam, column and shear wall, always having in mind inequality 1 and the fact that the axial (N), bending (M) and shear (Q) force demands in each one of these structural members from the combination of dead and live loads plus the earthquake forces are of a particular nature. Thus, for the slabs the demands are mainly flexural whereas for the beams the demands are flexural and shear. For the columns the demands are flexural and shear with the presence of considerable axial forces whereas for the shear walls the demands are flexural and shear with the presence of a relatively lower level of axial forces than that of the columns. Apart from the structural elements themselves, one should also consider critical areas of their connections (joints) as well as the foundation. Both, the structural connections and the foundation are very critical areas that require special consideration in both identifying the nature of the structural damage as well as proposing countermeasures. This presentation does not deal with either of these critical areas. The main flexural structural damage in slabs and beams develops in the areas of maximum bending moments. For the beams they usually develop near the joints with the columns and shear walls (figure 2a) where large bending moments are expected to develop from the seismic forces. Similarly, at the ends of the beams are the areas of large shear forces from the combination of earthquake forces with the dead and live loads; these will cause the appearance of shear damage in the form of diagonal cracks (figure 2a). The presence of large bending moments mainly from the seismic load together with large axial forces will cause the formation of flexural damage at the top and the toe of columns (see figure 2b) whereas the presence of shear forces from the seismic loads together with axial forces will lead to the formation of shear damage at the columns (see figure 2c). The presence of large shear forces from seismic loads together with relatively low level axial forces will lead to the development of shear damage in the shear walls (see figure 3a) whereas the presence of short columns will lead to the development of large shear forces from seismic loads and the development of shear damage as shown in figure 3b.

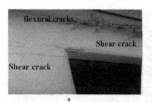

a   b   c

Figure 2. a Typical damage of beam at the joint with the nearby column (6th story building, Aharnes, Athens earthquake 1995), [2]. b Typical damage of the toe of a column at ground floor (two-story building, Nea Kifisia,, Athens earthquake 1995), [2]. c Typical shear damage at the columns of a "soft story" (4-story Metamorfosi building, Athens earthquake 1995), [2].

1.3. The strategy for the repair and strengthening scheme

In the previous discussion the emphasis in the description of damage was on attributing the various forms of structural damage to the nature of the demand (either flexural or shear with or without axial forces). However, as already stressed by inequality 1, structural damage is due to the fact that the said demand was not met by the existing strength. The strategy for retrofitting damaged structural systems, or structural systems that can be demonstrated by analytical methods prior to a strong earthquake to be prone to potential damage in the future, must be based on either somehow lowering the demands or increasing the corresponding strengths or both. In an effort to lower the demands, one can try to decrease the masses mobilized by the earthquake vibrations. This can be accomplished either by removing unnecessary mass or by changing the structural system in such a way that the resulting dynamic system combined with the design spectrum leads to a decrease in the earthquake loads (e.g. seismic isolation). Lowering the demands in the way described before is not always feasible. Thus, the retrofitting scheme is usually based on increasing the strengths of the structural members. In doing so, one must be aware that it is advisable to increase the deformability of the structural members thus increasing the ability of the structural system to dissipate the seismic energy through plastic deformations that are designed to develop at predetermined locations ([7], [8] [9], [10]). The location where these plastic deformations occur should be such that it does not lead to unstable structural formations. Sometimes, a partial objective of the strengthening scheme is to increase the stiffness to a moderate degree, especially for structural systems that can develop excessive torsional response. However, frequently a considerable increase in the stiffness of the strengthened structural system results as an indirect consequence of the adopted scheme whereas its main objectives were to increase the strength and deformability of the structural elements and of the structure as a whole. This increase in stiffness results in larger demands due to the dynamics of the strengthened structure, as it corresponds to higher amplitudes of design spectral accelerations than the un-strengthened structural system. This is usually the case when traditional strengthening schemes are employed utilizing reinforced concrete jacketing of the structural members (columns and beams) or the addition of shear walls by encasing reinforced concrete elements within the bays of the corresponding R/C frames [6]. The strengthening schemes that can be employed in order to increase the flexural or shear capacity of columns, beams or shear walls utilizing fiber reinforcing plastics (FRP) usually made of carbon, glass or steel do not result in this undesirable increase in stiffness. Moreover, due to their external application, they usually require less interference with or breaking of the volume of existing R/C structural elements. Finally, such strengthening schemes become effective in a much shorter time than traditional strengthening schemes ([11], [12], [13], [14], [15], [16]).

a   b

Figure 3. a Typical damage of shear walls at the "soft story" during The Kozani 1995 Earthquake in Greece. b. Typical damage of "short" columns during The Kozani 1995 Earthquake in Greece.

2. Main applications for dealing with earthquake structural damage utilizing fiber reinforcing plastics

In section 1.2 a brief description of earthquake structural damage that usually develops in slabs, beams, columns or shear walls of multi-story R/C buildings was presented. In this section, the use of fiber reinforcing plastics (FRP) will be discussed in a way dealing with the corresponding damage. These FRP materials behave in tension almost elastically till their ultimate state that for the material itself is the breaking of the fibers in tension; they do not develop any forces in shear or compression. The value for the modulus of elasticity is approximately 240Gpa for carbon fibers, 200Gpa for steel fibers and 80Gpa for glass fibers ([8], [17]).

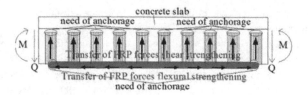

Figure 4. Flexural and shear FRP reinforcement for a T-beam.

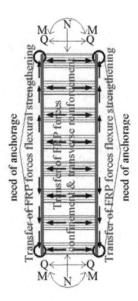

Figure 5. Flexural and shear FRP reinforcement for a column.

The ultimate axial strain values given by the manufacturers reach values in the range of 2%. Consequently, sheets made by these materials, despite their relatively small thickness which is usually below 0.2mm for one layer, can develop substantial tensile forces in the direction of their fibers (see figures 4 and 5). This property accompanied with their low weight and the very easy external application to structural elements, by attaching them on the external surfaces by proper organic or inorganic matrices, results in their being used as effective longitudinal or transverse reinforcement for such structural elements that are in need of strengthening (see figures 4, 5 and [12], [15], [16]). However, the following limitations exist for this type of application. The first limitation springs from the fact that the ultimate axial strain value of the order of 2% for the material of the fiber cannot be reached for all the fibers together in a sheet due to the actual conditions of the attachment that results in non-uniform distribution of the strain field ([13], [18], [19]). Thus, for design purposes, the adopted limit axial strain value is usually below ½ of the ultimate axial strain value for the fiber material. The second limitation results from the way the tensile forces which develop on these FRP sheets can be transferred. When the transfer of these forces relies solely on the interface between the FRP sheet and the external surface of the reinforced concrete structural elements, the delamination (debonding) mode of failure of these sheets occurs, due to the relatively low value of either the ultimate bond stress at this interface or the relatively low value of the tensile strength of the underlying concrete volume. This mode of failure is quite common and it occurs in many applications well before the corresponding FRP sheets develop tensile axial strains in the neighborhood of values mentioned before as design limit axial strains (approximately of the order of 1%). Consequently, there is need of alternative ways in order to transfer these tensile forces apart from the simple attachment, in order to enhance the exploitation of the FRP material potential. This will be presented and discussed in the following paragraphs ([11], [13], [20], [21], [22], [23]).

2.1. Upgrading the flexural capacity of slabs

The main cause of damage in this case is the fact that the flexural capacity cannot meet the demand. Fiber reinforced plastics either in the form of sheets or laminates are externally attached to the slab either at the top or at the bottom surface (positive or negative bending moment demand). Such a scheme was utilized in the upgrading of the seismic capacity of the slabs of a 4-story R/C building built in 1933 and subjected to the Kozani, 1995 earthquake in Greece ([2], [24]). For this strengthening scheme it was planned to make use of a certain type of Carbon Fiber Reinforcing Plastic laminates (CFRP) with a cross-section 50mm x 1.2mm. These laminates could be applied in-situ either on the upper or lower surface of the concrete slabs with the use of a special epoxy paste. An extensive experimental parametric study was performed on this type of attachment by testing a series of specimens prior to applying the best attachment detail to prototype slab specimens. The tests were performed at the laboratory of Strength of Materials of Aristotle University and utilized twin prismatic concrete specimens with dimensions 100mm x 100mm x 150mm each. These twin concrete prisms were joined with these CFRP laminates; they were attached on the two opposite sides of each twin concrete prism. These specimens were then subjected to such a loading as would force the laminates to be detached from the concrete surface. This is shown in figures 6a to 6d. A variety of attachments between the CFRP laminates and the twin concrete prisms were tried. The simplest form of attachment was the one employing only epoxy paste between the CFRP laminates and the concrete surface (figures 6a and 6b). Next, a variety of bolting arrangements were utilized with or without the epoxy paste. Figures 6c and 6d depict such an attachment of the laminates whereby the epoxy paste is combined with one bolt on each side of the twin concrete prism. The bond strength in the first case was found equal to 2.70Mpa whereas in the second case equal to 4.40Mpa, which represents an increase of 63%. The maximum bond strength that was achieved throughout these tests reached the value of 7.20Mpa, which represents an increase of 167% from the simple attachment of the laminates only with epoxy paste. These findings were utilized with slab specimens that were taken from the actual mezzanine slab of the prototype structure; they were cut from parts of the slab where an opening would be formed for a new staircase.

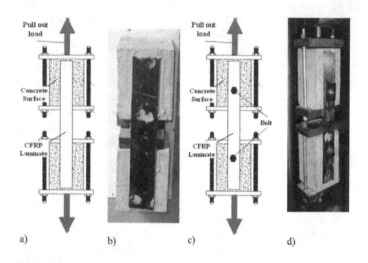

Figure 6. Bond tests of the used CFRP laminates

Figure 7. Loading arrangement of the slab specimens in flexure.

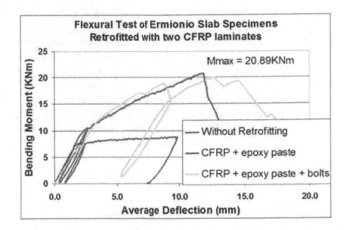

Figure 8. Flexural behavior of Ermionio slab specimen retrofitted with two CFRP laminates either bonded only with epoxy paste or with epoxy paste and bolts.

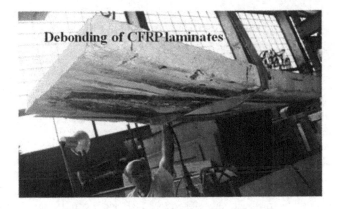

Figure 9. Flexural failure of the Ermionio slab specimen retrofitted with two CFRP laminates bonded only with epoxy paste. The CFRP laminates were debonded from the concrete slab in this case.

Figure 7 depicts the loading arrangement that was used to subject these slab specimens to four-point flexure. Initially, this was done for a specimen without any retrofitting that reached a maximum bending moment value equal to 8.84KNm. Then two laminate strips, each having a cross section of 50mm x 1.2mm, were attached on the bottom surface of this slab specimen, which was reloaded to flexure as before. Figure 8 depicts the obtained flexural behavior of this retrofitted slab specimen. At this stage, the CFRP laminates were attached to the specimen only with the use of epoxy paste. This time the specimen reached a maximum bending moment value equal to 20.89KNm, which represents an increase equal to

136% when compared to the maximum bending moment value of the un-retrofitted slab. This retrofitted slab failed in the form of the debonding of the CFRP laminates (figure 9). The debonding of the two CFRP laminates was followed by a sharp decrease in the flexural bearing capacity together with the formation of a plastic hinge at mid-span in the form of excessive plastic deformations of the old steel reinforcing bars at the bottom of the slab as well as crushing of the concrete at the top of the slab. Next, two new CFRP laminates were reattached to the same slab specimen; this time epoxy paste was used again together with bolts penetrating through the slab and securing the attachment of these CFRP laminates. The flexural bearing capacity of the slab this time reached the same maximum bending moment value as the one observed when only epoxy paste was used for the attachment of the laminates; they exhibited satisfactory performance without any signs of failure either in the form of debonding or fracture. This is shown by the bending moment versus mid-span deflection curve plotted for this test in figure 8. Despite the satisfactory performance of the CFRP laminates, no increase in the flexural bending capacity could be achieved due to the slab damage from its previous loading history; however; the performance of the anchors for the attachment of the CFRP laminates was satisfactory and the flexural behavior of the slab exhibited a much larger range of deformability (figure 8).

2.2. Upgrading the flexural and shear capacity of beams

The upgrading of the shear / flexural capacity will be presented in this section together with the corresponding modes of failure of reinforced concrete (R/C) beam specimens including repair schemes with Fibre Reinforcing Plastics (FRP) ([13], [25], [26], [27], [14], [28], [29]). The specimens were constructed and tested at the laboratory of Strength of Materials and Structures of Aristotle University. The applied load and the deflections of the specimen at three points were monitored throughout the experiments. Moreover, strain gauges were applied at selective locations of the FRP strips employed in the repaired specimens in order to monitor their state of stress during the loading sequence. In what follows, the most important experimental results are presented together with a discussion of the observed performance in order to highlight the role of the FRP strips in the observed behavior. It must be stressed that in all the studied beam specimens closed loop FRP hoops could be applied; this cannot be done in T-beam sections [19].

2.2.1. The initial reinforced concrete beam specimen and its observed behavior.

The virgin beam specimen: The virgin specimen BEAM-1 is a typical R/C beam, shown in figure 10, of rectangular cross-section of 200 mm x 500 mm and length l = 3700mm, having longitudinal reinforcement 3Φ20 at the top and 3Φ20 at the bottom side with closed stirrups as transverse reinforcement Φ8/250 ([14]). The selected longitudinal and transverse reinforcement together with the loading arrangement, shown in figure 11, will lead the behavior of this virgin specimen to be dominated by the shear rather than the flexural mode of failure.

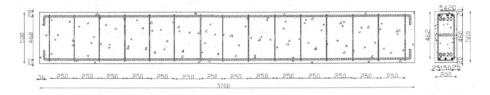

Figure 10. Dimensions (in mm) and structural details of specimen *Beam-1*. Longitudinal reinforcement: 3Φ20 at the top and 3Φ20 at the bottom. Transverse reinforcement: Φ8/250.

The behavior of BEAM-1: The specimen was simply supported at two point supports and was subjected to loads at two points symmetrically spaced at a distance 1150mm from each support; the mid-part had a length of 700mm. This loading was gradually increased, being monitored all the time. Moreover, throughout the loading sequence till the failure of the specimen, the cracking pattern was monitored together with the deflections of the beam at three points (mid-span and under each point load). As can be seen in figure 11, the shear cracking pattern is predominant at the two sides of the beam, whereas the middle part, as expected, is dominated by the flexural behavior. Finally, the dominant failure mode was from shear, as was planned during the design of this specimen. Figure 12a depicts the observed shear cracking patterns. In order to facilitate the repair of this specimen the loading was stopped after a decrease in the bearing capacity of the specimen was observed. In this way, the permanent deformations of this specimen at this stage were rather limited and the shear mode of failure was not allowed to fully develop.

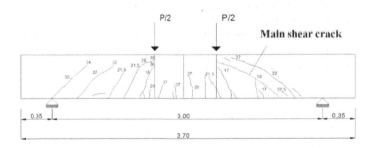

Figure 11. Flexural and shear cracking patterns for specimen *Beam-1*. Prevailing final mode of failure the shear.

Evaluation of the observed behavior for specimen BEAM-1: Figure 12b depicts the observed behavior in terms of shear force (Q). In the same figure the horizontal lines of different colours indicate the shear capacity of this particular specimen, as was predicted following the procedure suggested by various researchers ([30], [31], [32]) as well as by either the Greek Code [9] (without safety factor) or by the Euro-code [7] (with safety factor). Figure 12c depicts the observed behavior in terms of bending moment (M). Again, the horizontal lines of different colours in this figure indicate the flexural capacity of this specimen as predicted by well established formulae as well as by the corresponding code procedures (without safety factor). Bearing in mind that the mode of failure is dominated by shear and not flexure the comparison between the shear and flexural limit states predicted in this way confirms the observed behavior.

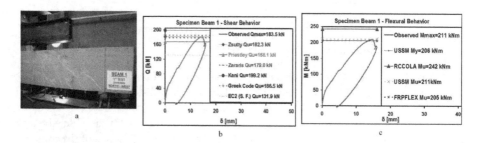

Figure 12. a Cracking patterns of Beam-1. b Evaluation of Shear Behavior (BEAM-1). c Evaluation of Flexural Behavior (BEAM-1).

2.2.2. Repaired beam specimen BEAM 1R-1 and observed behavior.

Description of Specimen BEAM 1R-1: After the end of the loading sequence of specimen BEAM-1, this specimen was repaired against shear failure in the way shown in figure 13. Closed loop CFRP (type C1 – 23) hoops of two layers (of nominal thickness 0.133mm per layer), having width b_{slice} = 75mm, were attached externally on the specimen spaced at distances of s=275mm. This was very easy to apply to this specimen and the only necessary treatment was to round up the corners of the specimen. However, in order to apply exactly such a repair scheme in prototype conditions one is faced with the considerable difficulty in drilling the appropriate holes at the top R/C slab that is usually monolithically connected to the beam at its top part. Moreover, the attachment of the CFRP layers in closed loop hoops will cause considerable difficulty. Alternatively, the CFRP shear reinforcement may not be of the closed loop hoop type but of the open loop U-shaped strips with or without anchors. The shear and flexural cracking that developed in specimen Beam-1 from the previous loading sequence was left without any repair (by epoxy resins or other means). An external layer of resin was used to paint the areas of the cracking in order to be able to monitor the activation of these cracks during the new loading sequence [14]. The mechanical properties of the employed CFRP layers (type C1 – 23), as they are given by the manufacturer, are : Ultimate stress f_u = 3800 Mpa, Young's Modulus E_f = 230 Gpa, Limit strain ε_u = 1.8 %, thickness t_f = 0.133 mm.

The Use of Fiber Reinforced Plastic for The Repair and Strengthening of Existing Reinforced Concrete
Structural Elements Damaged by Earthquakes

103

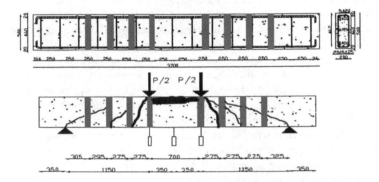

Figure 13. Repaired specimen BEAM 1R-1 The newly activated cracks are shown with bold lines. Dominant mode of failure the one at the compression zone of the mid-part of the specimen governed by flexure.

Observed Behavior of Specimen BEAM 1R-1: The main shear crack at the right side gave signs of becoming active for vertical load of 177KN (Q=88.5KN) whereas the same occurred for the left side main shear crack for vertical load of 216KN (Q=108KN). This meant that the corresponding CFRP shear type hoops started to develop stresses resisting part of the shear. Crushing was observed at the compressive upper side of the mid-part of this specimen undergoing flexure, when the vertical load reached 373KN. This was accompanied by the activation of the flexural cracks starting from the bottom side of the beam. These vertical flexural cracks developed even further for a maximum load of 414KN extending all the way from the bottom side to the crushed upper side, with the specimen reaching its flexural capacity. The shear mode of failure of the previous specimen was contained without further development by the applied CFRP hoops. The final condition of this specimen is depicted in figure 13. The observed behavior in terms of shear force (Q) and bending moment (M) is depicted in figures 14a and 14b. The maximum observed shear force was Q = 207 KN, and the maximum observed bending moment was M= 238 KNm.

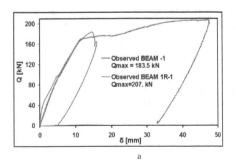

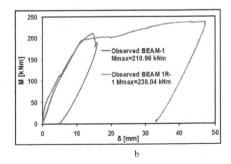

Figure 14. Observed shear behavior of BEAM 1R-1. 14b. Observed flexure of BEAM 1R-1.

In these figures 14a and 14b, the observed behavior before the repair of this specimen is also plotted for comparison with maximum observed shear force Q = 183.5 KN, and maximum observed bending moment M= 211 KNm. From this comparison it can be concluded that the applied CFRP hoops resisted a shear force of 23.5KN without signs of any distress. The successful increase (13%) in the shear bearing capacity of this specimen by this applied CFRP repair scheme, which leads to the exhausting of the specimen's flexural capacity and to reaching the desired flexural mode of failure, demonstrates the potential of such repair schemes for prototype conditions. However, there are construction difficulties in prototype conditions which were already pointed out before. Moreover, as was remarked before, the shear capacity of this scheme was not reached as the specimen survived the applied loading sequence in terms of shear. A subsequent reloading of specimen BEAM 1R - 1, with additional repair elements, namely specimen BEAM 2R -1, is presented in the following. The objective is to define by measurements the shear bearing capacity of the applied CFRP-hoops in this repair scheme.

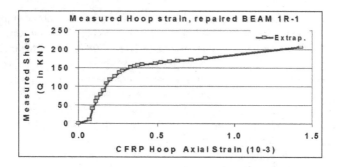

Figure 15. The variation of the CFRP-Hoop strain with the measured shear force for BEAM 1R-1.

Evaluation of the contribution of the CFRP for specimen BEAM 1R-1: During the loading sequence of Beam 1R-1, described above, the strains that developed at the mid-height of the third CFRP-hoop from the right end of the specimen (see figure 13) were monitored throughout the test by a strain gauge attached in this location. This strain gauge remained operational up to a total vertical load of 353KN (Q=177.6KN) measuring for this load an axial strain equal to 0.081 %. Unfortunately, beyond this load, the strain gauge became inactive due to a crack going through it. The variation of the axial (vertical) strain of CFRP-hoop with the applied shear force is depicted in figure 15. As can be seen in this figure, during the loading of specimen Beam 1R-1 the variation of the measured CFRP-hoop strain with the applied shear force is almost linear with different slope in two distinct ranges; the first range for shear force values up to Q=150KN and the second range for shear force values larger than Q=150KN. By extrapolating the second linear strain trend up to the maximum measured shear force value of 207KN a maximum strain value equal to $1.43 \cdot 10^{-3}$ is found for the CFRP-hoop, which is also depicted in figure 15 (see also figure 14a). By employing the relationship: $V_f = A^{hoop}(h\,/s)E_f\,\varepsilon_f\,{}^{eff}$

where $A^{hoop}=39.9mm2$ the cross-section of one CFRP hoop, $h=500mm$ the height of the beam, $s=275mm$ the spacing of the hoops, $E_f=230Gpa$ the Young modulus of the applied CFRP, $\varepsilon_f{}^{eff}$ $=1.43\ 10^{-3}$ the effective CFRP strain as was mentioned before, the value of the shear resisted by the applied CFRP-hoops can be found equal to $V_f=23.86KN$.

When this value, which was found from direct strain measurements of the CFRP-hoop, is compared with the value of $V_{fa}=23.5KN$, found by subtracting the shear resisted by the virgin specimen BEAM-1 from the shear resisted by the repaired specimen BEAM 1R-1 with the CFRP-hoops, a very good agreement is observed, which further validates the shear resisted by the applied CFRP-hoops. Using the software developed by Triantafillou [33] a prediction for the maximum shear force equal to 44KN is found to be resisted by the CFRP-hoop arrangement applied in BEAM 1R-1 (figure 13). The measured value of approximately 24KN and the observed shear performance of this specimen are in agreement with this predicted upper limit (44KN) for the CFRP-hoop shear resistance of the employed repaired scheme.

2.3. Repaired specimen Beam 1R-2

As described in section 2.2.2. before, a middle part compressive zone crushing was observed as a limit state for specimen BEAM 1R-1. After unloading this specimen the crushed concrete was removed and the whole area cleaned till the compressive longitudinal reinforcement became visible. The reshaping of the orthogonal form of this part of the beam was achieved with a paste from a mixture of cement and epoxy resin, which becomes hard in one day attaining a compressive strength of 30 MPa. The thickness of this layer of paste was 20mm above the compressive longitudinal reinforcement and the corners were given a smooth curvature. This upper zone was next reinforced with layers of CFRP (C 1-23), as shown in figure 16, in order to increase its compression capability through confinement. In addition, three layers of GFRP (G-60 AR) were attached at the bottom side of the specimen to enhance through tension its flexural capacity beyond the one provided by the existing longitudinal reinforcement (figure 16).

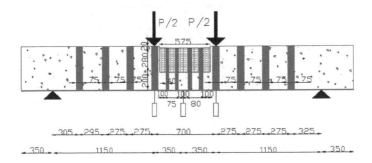

Figure 16. Specimen BEAM 1R-2 with CFRP layers attached in a way as to contain the compressive zone of the middle part as well as additional GFRP layer at the bottom side of the specimen extending the whole span.

2.3.1. Observed Behavior of Specimen BEAM 1R-2

This specimen was reloaded in a way similar to the one employed for the preceding specimens BEAM -1 and BEAM 1R-1. Figures 17a and 17b depict for all three specimens the shear and flexural behavior, respectively. In figure 17b the predictions of the yield moment for specimen BEAM-1 and the ultimate moment of specimen BEAM 2R-1 are also indicated. Due to an increase in the flexural capacity for specimen BEAM 2R-1 when compared to the preceding specimen BEAM 1R-1, resulting from the described repair scheme for specimen BEAM 2R-1, the shear failure mode prevailed this time. This mode of failure was accompanied by the rupture of the CFRP hoops as well as by crushing of the concrete at the compressive shear transfer zone neighboring the point of loading (figure 18). As expected, this specimen (BEAM 1R-2) exhibited an increase in its flexural capacity when this is compared with the flexural limit state attained during the preceding test of specimen (BEAM 1R-1). This fact demonstrates that the repair scheme described before, aimed to increase the flexural capacity of specimen BEAM 1R-2 proved to be successful. The observed maximum shear force during this test was Q_{av} = 215.8KN whereas the maximum bending moment was M_{av} = 248.2KNm. Comparing these values with the ones observed during testing the original virgin BEAM-1 it can be concluded that the applied repair scheme resulted in an increase of the order of 18% for the shear capacity and more than 18% for the flexural capacity; the exact increase in the flexural capacity could not be measured as the shear failure for BEAM 1R-2 prohibited this specimen from reaching its flexural limit state. The predicted flexural capacity of BEAM 2R-1 is listed in Table 1.

METHODOLOGY	MU [KN m]	Mode of failure
Ultimate stress strain method	**301.26**	Failure from compression of the top zone and of tension of the GFRP
FRPFLEX	**319.02**	Failure of compression zone

Table 1. Predictions of flexural capacity for BEAM 2R-1.

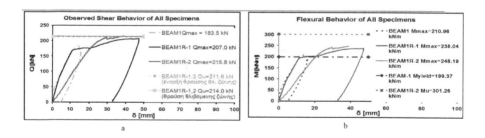

Figure 17. Shear behavior of all specimens. b. Flexural behavior of all specimens.

It must be pointed out here that the way the enhancement of the compressive zone capacity was achieved through the compressive zone confinement, attained from the CFRP layers, cannot be applied so easily in prototype conditions due to the presence of an R/C slab at this part of a prototype beam. Table 2 lists the total vertical load values which were measured during important developments in the observed behavior of specimen BEAM 1R-2. Moreover, figure 19 depicts the observed mode of failure in a detailed way.

P [KN]	Description of the observed cracking
73.6	Partial activation of the shear cracks that were formed from past loading
166.8	Activation of the previously formed flexural cracks
235.4	Formation of main flexural crack next to the CFRP stirrup at loading point (left side)
343.4	Formation of main flexural crack next to the CFRP stirrup at loading point (right side)
421.8	Initiation of crushing of the concrete compressive zone at the shear transfer area neighboring the left loading point.
423.8	Rupture of a CFRP stirrup
431.6	Crushing of the concrete at the compressive shear transfer zone neighboring the left loading point. Rupture of the CFRP hoops that were crossed by the main shear crack (figure 18).

Table 2. Important developments in the behavior of specimen BEAM 1R-2 during loading.

Figure 18. Shear failure accompanied by the rupture of the CFRP stirrups as well as of crushing of the concrete compressive zone at the shear transfer area neighboring the point of loading.

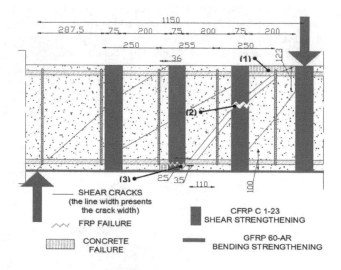

Figure 19. Mechanism of shear failure during the limit state of specimen BEAM 1R-2.

2.3.2. Evaluation of the contribution of the CFRP-hoops for BEAM 1R-2

As was done for specimen BEAM 1R-1, during the loading sequence of Beam 2R-1 the axial (vertical) strains that developed at the mid-height of the third CFRP hoop from the right end of the specimen (see figure 13) were again monitored employing a strain gauge attached in this location. This strain gauge remained operational throughout the whole load sequence measuring a maximum axial strain value $\varepsilon_f^{\,eff} = 0.67$ % when the total vertical load reached its maximum value of 435.6KN (Q=215.8KN). By subtracting the shear resisted by the virgin specimen BEAM -1 (Q=183.5KN) from the shear resisted by the repaired specimen BEAM 2R-1 (Q=215.8KN), a value of $V_{fa} = 32.3$KN is found as being resisted by the used repair scheme. As mentioned before, using the software developed by Triantafillou [33] the maximum shear force of 44KN is found to be resisted by the CFRP hoop arrangement applied in BEAM 2R-1. Moreover, employing the maximum strain measurement and the relationship $V_f = A^{\,hoop} (h/s) E_f \varepsilon_f^{\,eff}$ the shear resisted by the applied CFRP-hoops can be found equal to $V_f = 111.8KN$. This discrepancy dictates further investigation on the shear resisting mechanisms that developed during the limit state of specimen BEAM 2R-1 (see figure 19).

The maximum axial strain measurement equal to 0.67 % was recorded at this particular CFRP-hoop location during maximum applied load just before the rupture of this CFRP-hoop and the final development of the shear failure for specimen BEAM 2R-1. The variation of the applied shear force with the measured CFRP hoop strain is depicted in figure 20; it is almost linear with different slope in two distinct ranges; the first for a shear force value from zero to Q=205KN and the second for a shear force value larger than Q=205KN up to $Q_{av} = 215.8KN$, when the rupture of the CFRP occurred.

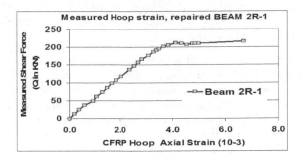

Figure 20. The variation of the CFRP-hoop strain with the measured shear force for BEAM 2R-1.

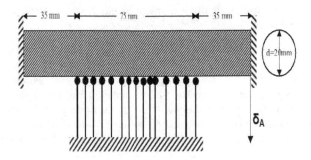

Figure 21. Numerical simulation of the dowel action.

2.3.3. Shear capacity of specimen BEAM 2R-1 based on the observed failure mechanism

Based on the observed shear failure mechanism, which is depicted in figure 19, the shear capacity of BEAM 2R-1 will be estimated as will be described below. It is assumed that the shear is resisted in this case by the following parts.

One closed steel stirrup.

The compressive zone for shear transfer next to the point of load application at the top of the main shear crack.

Two CFRP-hoops.

The contribution of the bottom side longitudinal reinforcement (dowel action).

The contribution of the compressive zone is found from the relationship: $V_c = 0.3 \ (fc)^{2/3} \ (2)^{1/2}$ x b. Where fc the concrete compressive strength, x the height of the compressive zone (as estimated from the flexural behavior), b the width of the beam. This leads to a shear force equal to $V_c = 46.42$ kN at the initiation of concrete compressive zone crushing.

The contribution of the assumed two CFRP-hoops is equal to $V_f = 18234 \, \varepsilon_f^{\, eff}$ [KN]. According to the strain measurements this contribution at the initiation of crushing of the compressive zone is equal to $V_f = 73.54$KN.

The contribution of the dowel action of the longitudinal reinforcement will be based on the geometry of this particular failure mechanism which was measured in detail at the laboratory. For this purpose, a simple numerical simulation of this dowel action was formed, which is depicted in figure 21. The longitudinal reinforcement is supported by a steel stirrup (left support) and by the CFRP-hoop (multiple mid-supports). All the supports are simulated by springs with the corresponding elastic / post-elastic and cross-sectional properties the actual steel stirrup and CFRP-hoop possess.

Due to the volume of concrete the rotations at the two ends of this dowel action numerical model were restrained; moreover, the employed simulation provided the capability for the longitudinal reinforcement to develop plastic hinges with parameters based on the cross-sectional area and yield stress of the employed longitudinal reinforcement, which were measured at the laboratory.

The limit state contribution of the dowel action of the longitudinal reinforcement, as resulted from this numerical simulation, was found equal to $V_d = 37.67$ KN. The contribution of the steel stirrup (V_w) is based on the assumption that its strain is the same as the measured strain of the neighboring CFRP hoop. The various contributions based on all the above assumptions are listed in Table 3 together with the corresponding total measured value. As can be seen a very good comparison is reached between the total estimated shear resistance and the corresponding measured value. It must be pointed out that the estimated value was based on the exact observation of the failure mechanism as well as on the measurements of the CFRP hoop strains. In the same table a predicted shear resistance value is also listed based on the prediction of the maximum shear resisted by the CFRP-hoops by the software developed by Triantafillou [33] (maximum predicted V_f value equal to 44KN is found).

Steel Stirrup V_w [KN]	2 CFRP Hoops V_f [KN]	Dowel action of longitudinal reinforcement (3Φ20) V_d [KN]	Compressive Zone of Concrete V_c [KN]	Total estimated shear resistance V_R [KN]
53.94	73.54	37.67	46.42	211.57
Measured total shear resistance **210.9KN**			Predicted total shear **227.5KN** [ref. 33]	

Table 3. Shear resistance contributions of the various parts of specimen BEAM 2R-1 at the initiation of the compressive zone crushing.

The variation of the estimated shear contributions of the CFRP-hoops (Vf), the steel stirrup (Vw), the dowel action of the longitudinal reinforcement (Vd) as well as the sum of all these (Vf+Vw+Vd) are plotted in figure 22 together with the measured total shear force (V_R) for specimen BEAM 2R-1. The common basis for the plotted displacements between the estimated contributions and the measured values is found from the correspondence of the load and strain measurements during testing. In addition, in the same figure, the variation of the

contribution of the compressive concrete zone is plotted by subtracting from the total measured shear force (V_R) the contributions of the steel stirrups and the CFRP hoops (Vc=V_R-Vw-
Vf). After the formation of the shear cracks this contribution includes the dowel action of the
longitudinal reinforcement. This results in a maximum value Vc = 84.14 KN, which is in
agreement with predictions based on formulas suggested by various researchers. Moreover, it can be seen that the summation of the various contributions is in good agreement
with the total measured shear force if the contribution of the compressive zone is added, after
the formation of the shear cracks and up to the initiation of the crushing of the compressive zone concrete.

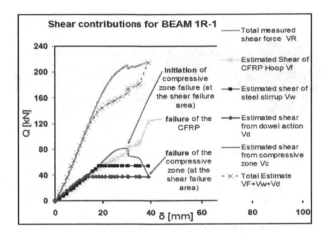

Figure 22. Shear resistance contributions, extracted from the mechanism of shear failure for specimen BEAM 2R-1.

2.4. Concluding observations

a1. The technique of repairing reinforced concrete beam elements by attaching CFRP layers
has been examined in the laboratory. It was found that using CFRP-hoops, spaced at the
areas of the beam that resist the shear forces, was a quite successful repair technique in upgrading the shear capacity of beam specimens by inhibiting the shear mode of failure thus
favouring the development of the flexural mode of failure.

b1. The repairing of the examined specimens by attaching GFRP layers to complement the
tensile longitudinal reinforcement together with CFRP confinement of the compressive zone
at these parts of the beam that undergo flexure was again a quite successful technique in upgrading the flexural capacity of the beam specimens. However, this must be properly controlled in order to prohibit the development of the undesired shear mode of failure.

c1. In order to be able to successfully control the shear and flexural capacities and modes of
failure, after the application of CFRP repair schemes, reliable tools are necessary that successfully predict the bearing capacities of the repaired structural elements. It was demon-

strated by this work that whereas this is possible for the flexural capacity it is not so straightforward for the shear capacity unless the basis of the shear capacity estimate is founded on a realistic failure mechanism that takes into account the realistic contributions of all the parts that are mobilized through this failure mechanism.

d1. It must be pointed out that the examined repair schemes for upgrading the shear resistance through CFRP stirrups will not be so easily applied in prototype conditions where the presence of a reinforced concrete slab, cast monolithically with the upper part of the beam, must be confronted.

Figure 23. R/C cross-sections with one side longer than the other (h/b>1.5) and CFRP partial confinement.

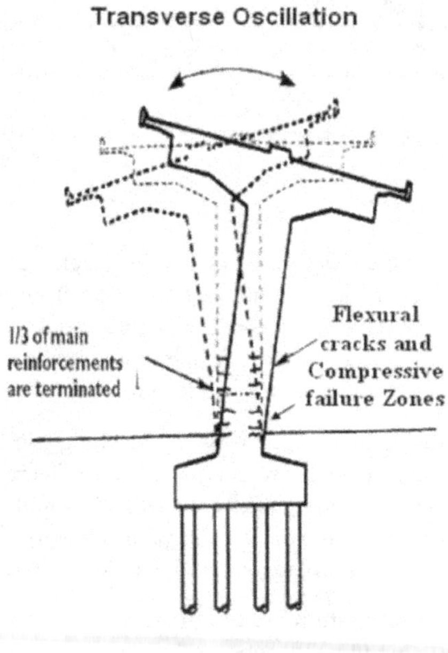

Transverse Oscillation

1/3 of main reinforcements are terminated

Flexural cracks and Compressive failure Zones

Figure 24. Bridge Pier compression failure mode.

3. Partial Confinement

3.1. The use of confinement for upgrading the flexural and shear capacity of columns.

A very common application of externally applied fiber reinforced plastics (FRP) is confining parts of structural members. This can be easily envisaged for columns with their cross-section approximating a square shape [34]. This is done by wrapping around the structural element one or more that one layers of FRP sheets together with the proper resine in-order to bond them on the surface of the structural member as well as to bond the FRP sheets between themselves. The main points of attention here are the following [35]:

Preparation of the surface of the existing structural member in order to prohibit premature delamination (debonding) of the attached FRP sheets.

Localized rupture of the FRP jacket due to abrupt change of curvature at the corners, to buckling of reinforcing bars, or to excessive dilation of concrete.

Proper application of the resin in order to achieve sufficient bond between the FRP sheets and the surface of the existing structural member.

Proper application of the resin and enough overlapping length of the wrapped layers of the FRP sheet around the existing member in order to prohibit any unwrapping mode of failure.

The proper wrapping of the FRP multilayer sheet around the existing structural member (column or beam) introduces a passive confinement in the same way as steel stirrups do on such concrete members. Whereas the wrapping of FRP sheets around columns or beams of rectangular shape with width / height ratio no greater than 2 can be easily materialized this is not so easily applicable in the following cases:

In columns with rectangular cross-sections having a width / height ratio greater than 2. It will be discussed later on in this section how this difficulty was confronted in the laboratory.

In T- beams whereby the presence of the slab prohibits the wrapping of the FRP sheets around the cross-section.

Grade	Compression failure mode	Increase in the compressive capacity (compared to the unconfined specimens)	Effectiveness of the partial confinement
1	Of the weak part as without the partial confinement	Small increase	low
2	Of the weak part as without the partial confinement	Considerable increase	Considerable
3	Of both the weak part and the strong parts	Substantial increase	Very effective

Table 4. Characterization of the effectiveness of the partial confinement.

The beneficial effect of such a confinement on the compressive strength of axially loaded structural members is well known and will not be repeated here. Moreover, the function of these externally applied FRP sheets as transverse reinforcement for both columns and beams is also evident in the same way closed steel stirrups function in reinforced concrete sections. Acting as external transverse reinforcement FRP sheets fully wrapped around beams or columns can be utilized to upgrade their shear capacity (figure 5). This has been already discussed in the beginning of section 2. Again, for T-beams the presence of the slab prohibits the formation of closed loop transverse reinforcement by the FRP sheets. If the FRP sheets are applied as transverse reinforcement in an open loop formation they are susceptible to a premature delamination type of failure (see figure 4). This will be further discussed in section 6 together with the way this difficulty was confronted in the laboratory.

The upgrading of reinforced concrete (R/C) cross-sections, with one side rather longer than the other (h/b > 1.5), by partial application of CFRP (Carbon Fiber Reinforcing Plastic) confinement is investigated here (figure 23). This partial application of CFRP confinement is aimed at the retrofitting of bridge-pier type R/C cross-sections in order to prohibit, up to a point, the development of premature compressive failure at the base of the pier due to combined compression and flexure from seismic loads (see figure 24 and [36], [37]). The performance of such structural elements was studied extensively in the past ([38], [39]). This type of partial confinement may also be applied to upgrading vertical structural members with non-accessible sides. Design guidelines for rectangular FRP jackets applied on rectangular columns have been proposed with the limitation that the cross-sections have aspect ratio h/b < 1.5 [37]. For higher aspect ratios it is recommended to design a circular or oval jacket. However, it is expected that for rectangular cross sections with aspect ratios larger than 1.5 the radius of a circular or rectangular jacket will be too large and will result in ineffectual confinement and will prove costly and impractical. For this reason it is desirable to investigate alternative schemes for increasing the confinement of rectangular cross-sections with relatively large aspect ratio without resorting to complete circular or oval jackets. Such a scheme is studied here using CFRP layers that do not extend all around the cross-section (figure 23 *"partial confinement"*).

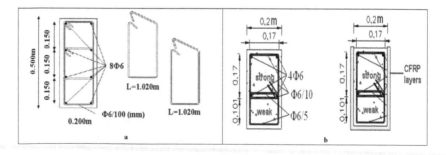

Figure 25. Initial cross-section with h/b ratio = 2.5. 25b. Test specimens without and with partial CFRP confinement.

To compensate for the fact that the CFRP layers do not enclose the cross-section entirely, anchorage of these layers must be provided, as shown schematically in figure 23. To this end, a laboratory investigation was carried out to study the effectiveness of such partial confinement together with alternative anchorage schemes. As will be explained, this effectiveness was tested by subjecting the specimens only to compressive loads. Despite this limitation, as will be demonstrated from the results of this investigation, the most significant aspects of the critical factors for this *"partial confinement"* scheme were brought to light.

3.1.1. The formation of test specimens

The initial cross-section, which formed the basis of the tested specimens, is depicted in figure 25a with an aspect ratio h/b equal to 2.5. This is a rectangular cross section of a bridge pier model structure, which was tested both at the laboratory (figure 25a) and at the Volvi-Greece European Test Site in the framework of the European project Euro-Risk [36]. This cross section was intentionally designed to develop flexural mode of failure at the base of the pier; moreover, it was desirable to find ways to retrofit such specimens by prohibiting premature compression failure at the base by means of partial CFRP confinement. The effectiveness of the partial CFRP confinement is studied by subjecting the tested specimens only to pure compressive loads. This type of stress field is expected to develop at the base of such vertical members under combined vertical loads and seismic actions, where undesired compression failure may develop. In order to limit the maximum level of compressive loads required to bring to failure such a cross-section with the loading arrangements available, the tested specimens had a cross section (figure 25b) of 200mm by 300mm instead of 200mm by 500mm of the initial cross-section (figure 25a and figure 25b) for the bridge pier specimens tested both at the laboratory and at the test site under combined compression and flexure [36]. Moreover, in order for the tested specimens to form compression failure at the same part of the cross-section where such failure would develop at the base of the initial bridge pier model, one part of the tested cross-section was left identical to the initial cross-section (the one that is marked in figure 25b as weak) whereas the remaining part was strengthened both with longitudinal and, in particular, with transverse reinforcement (the one marked in figure 25b as strong).

In this way, with the compression capacity of the weak part being smaller than that of the strong part, the compression failure was expected to develop at the weak part. This proved to be correct during the experiments, as will be shown in the following sections. The CFRP partial confinement was applied at the weak part, as is shown at the right hand side of figure 25b. Then, by studying the resulting bearing capacity and mode of failure under compression of the tested specimens (with or without partial confinement) the effectiveness of such a repair scheme could be demonstrated and classified as listed in table 4.

3.1.2. Construction of test specimens

Ten identical specimens were constructed and eight of them were used in the current experimental sequence (see table 5). All specimens were reinforced in the same way and were cast at the same time with the same mixture aiming for similar plain concrete strength values.

The reinforcing details are depicted in figure 26. As can be seen, the total height of the specimens was 1600mm. They had a mid-part height of 580mm that was left to develop the compression failure (figure 26, part denoted as CROSS SECTION A). The cross section of this mid-part was the one shown in figure 25b, including the two distinct parts (the weak and the strong). The two edges of the specimens, with a height of 510mm each, were confined during the experiment with strong steel brackets covering these parts from all sides thus prohibiting any compressive failure developing at those two edges (see figure 27).

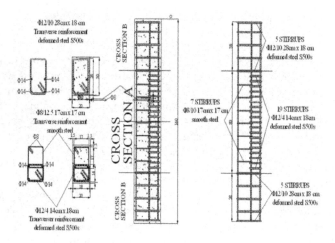

Figure 26. Reinforcement details common to all specimens.

Virgin Specimens	CFRP Confinement	Repaired Specimens	Plain Concrete Strength (Mpa)
Specimen 1 Test 1	No	Test 2 3+(2) CFRP	28.0
Specimen 1a Test 1	3 CFRP layers	Test 2 3 GFRP	25.8
Specimen 3 Test 1	No	Test 2 3+(2) CFRP	27.6
Specimen 3a Test 1	5 CFRP layers	Test 2 5 CFRP	27.6
Specimen 4 Test 1	5 CFRP layers		27.7
Specimen 4a Test 1	5 CFRP layers		27.7
Specimen 5 Test 1	5+(2) CFRP layers	Test 2, 5+2 CFRP Test 3, 5+2 CFRP Test 4 7 CFRP	27.6
Specimen 5a Test 1	No	Test 2, 7 CFRP Test 3 7+4 CFRP	25.8

Table 5. Test specimens with their corresponding concrete strength.

The partial CFRP confinement was attached only on the three sides of these specimens covering the weak part and leaving the fourth side (of the strong part) free without any CFRP layers (figure 25b). There were 8 virgin specimens, namely 1, 1a, 3, 3a, 4, 4a, 5, and 5a (Table 5). These specimens were tested in their virgin state in which some of them were without partial CFRP confinement whereas the rest had the partial CFRP confinement applied to them from the beginning. The second column of table 5 indicates the partial confinement condition of the virgin specimens. The testing sequence of these virgin specimens is signified as Test 1. Most of these specimens were repaired after they had reached their limit state during their previous test. In all the repaired specimens the CFRP partial confinement was applied. Throughout these series of experiments the nominal thickness of the employed CFRP layers was 0.176mm with a given Young's Modulus E=350GPa. The measured maximum axial CFRP strain was approximately 1%.

Figure 27. Specimen with confining steel brackets at the edges and partial CFRP confinement at the mid-part

This sequential test number of the repaired specimens is signified as Test 2 (for the 1ˢᵗ re-pair), Test 3 (for the 2ⁿᵈ repair) etc. (see column 3 of Table 5). In the same column the num-ber of CFRP layers used in the partial confinement for the repaired specimens is also indicated. The total number of specimens, virgin and repaired, was seventeen. In Table 5 the unconfined concrete compressive stress is also listed, found from cylinders with diameter 150mm and 300mm height; these cylinders were obtained during the casting of each virgin specimen. The anchorage of the CFRP layers was applied along the two long sides, which were common to both the weak and the strong parts. The main load that was applied was axial compression, although in limited specimens the axial compression was combined with bending, which is not reported here. From the observed behavior, the effectiveness of the applied partial confinement could be deduced. As shown in table 4, this judgment was based on the level of the bearing capacity combined with the type of compression failure that was formed (at the weak or strong part). Moreover, the observed behavior of the vari-ous parts of the test specimens, such as the CFRP layers and their anchorage, helped to iden-tify the factors that bear an adverse or beneficial influence on these aspects of the behavior.

3.1.3. Critical parameters and their variation:

The following parameters were critical (see results contained in Table 6).

1 ˢᵗ. The type of anchorage of the CFRP layers, 2 ⁿᵈ. The number of the CFRP layers.

The second parameter becomes critical only if the first parameter performs satisfactorily. A large number of specimens (1, 1a, 3, 3a, 4 and 4a) exhibited failure of the anchorage of the partial confinement thus limiting the effectiveness of the partial confinement to low levels. Consequently, the increase in the compressive bearing capacity in this case was relatively modest. The compressive load was monitored by continuous sampling throughout the test. The "Average Stress at failure" listed in table 6 is found by dividing the compressive load by the gross cross-sectional area of each specimen. Listed in table 6 are certain particulars of the anchorage of the CFRP layers. Anchors-1 signifies bolts of 7mm diameter and 60mm length that do not penetrate the cross-section in its whole width, whereas Anchors-2 signifies bolts of 10mm diameter that penetrate the whole section (220mm length).For specimens 5 and 5a, due to the improvement of the anchoring details of the partial confinement, the increase in the compressive bearing capacity of these specimens was 50% higher than the observed bearing capacity of similar specimens with no partial CFRP confinement. When the maxi-mum compressive bearing capacity achieved with the partial confinement is compared with the corresponding plain concrete strength value, a 230% increase can be observed. More-over, the compressive failure in this case involved both the weak and the strong part thus improving substantially the effectiveness of the partial confinement and allowing the em-ployment of a large number of CFRP layers.

Specimen No	Virgin	Repaired	Number of CFRP layers	Anchors-1	Anchors-2 + washers	Average Stress at failure (Mpa)	Failure Mode
1 Test 1	Yes	No	No	No	No	41.12 (146.9%)*	Weak part
1 Test 2	No	Yes/ EMACO	3 (+2)	Yes	yes	41.69 (148.9%)	Bolts
1a Test 1	Yes	No	3	Yes		45.78 (177.4%)	Pull out Anchors-1
3 Test 1	Yes	No	No	No		40.79 (147.7%)	Weak part
3 Test 2	No	Yes	3 (+2)	Yes	Yes	45.78 (165.9%)	anchorage
3a Test 1	Yes	No	5	Yes	Yes	47.09 (170.6%)	anchorage
3a Test 2	No	Yes/ EMACO	5	Yes	Yes	42.51 (154.0%)	anchorage
4 Test 1	Yes	No	5	No	Yes weak	46.60 (168.2%)	anchorage
4a Test 1	Yes	No	5	No	Yes weak	45.53 (164.4%)	anchorage
5 Test 1	Yes	No	5 (+2)	No	Yes strong	53.96 (195.5%)	Steel bracket
5 Test 2	Yes	-	5 (+2)	No	Yes strong	55.26 (200.2%)	Steel bracket
5 Test 3	-	Top / EMAKO	5 (+2)	No	Yes Strong	53.96 (195.5%)	CFRP mid-height
5 Test 4	No	Yes/ EMAKO	7	No	Yes Strong	58.86 (213.3)	Strong stirrups
5a Test 1 bending	Yes	No	No	No	No	40.88 (158.4%)	Weak part
5a Test 2	No	Yes	7	No	Yes Strong	57.23 (221.8%)	CFRP mid-height
5a Test 3	No	Yes	7 +4	No	Yes Strong	60.17 (233.2%)	Strong part

Table 6. *Summary of test results together with the basic specimen characteristics.* *As % of the corresponding plain concrete strength.*

3.1.4. Instrumentation to obtain the average stress-strain behavior

Apart from monitoring the compressive load, the deformations of the mid-part were also continuously recorded throughout each experiment with displacement measurements taken at each side of the cross-section. Eight displacement sensors (two at each side) were employed to record the deformations of the mid-part. Although the deformation of this mid-part was far from uniform, as can be seen from the obtained displacement measurements of the weak and strong parts (figure 29), the average axial displacement, which was found by averaging the measured displacement values at all four sides of each specimen, is mostly used here as an indication of the deformability of each specimen. By dividing this average axial displacement by the height of the mid-part an average axial strain could also be obtained in this way. The following discussion of the observed behavior of each specimen is based on diagrams of average axial stress versus average axial strain found from the previously described averaging process. More detailed study on the obtained non-uniform deformability for each specimen will be carried out at a future stage. An additional measurement that was obtained during the experimental sequence was the axial strain that developed at the CFRP layers of the partial confinement of the mid-part. These CFRP strain measurements are an additional indication of the effectiveness of the partial confinement.

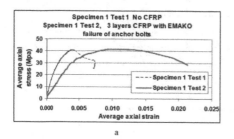
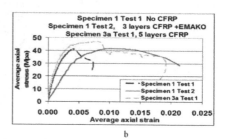

a   b

Figure 28. a Partial confinement of low effectiveness. b. Partial confinement of low effectiveness.

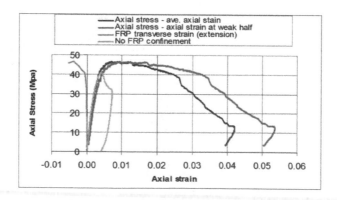

Figure 29. Specimen 4, Test 1 (5 layers CFRP) Comparison with Specimen 1 Test 1 (no confinement).

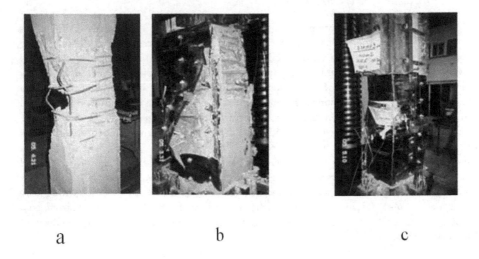

a   b   c

Figure 30. a Virgin Specimen without confinement. b. Failure of anchor bolts. 30c Low effectiveness of partial confinement. c. Tensile failure of CFRP Layers. Considerable effectiveness of partial confinement.

3.2. Discussion of the results

3.2.1. Partial confinement of low effectiveness

In figures 28a and 28b the obtained behavior of specimen1 (Test 2) with 3 CFRP layers and
specimen 3a (Test 1) with 5 CFRP layers is compared with specimen 1 test 1 (no partial confinement). Specimen 1 test 2 was formed from specimen 1 test 1 by repairing the failed specimen 1 test 1 with special (low shrinkage) concrete as well as with 3 layers of CFRP forming
the partial confinement. This repaired specimen failed in compression with almost the same
capacity as the previously-tested virgin test with no partial confinement, but with larger deformability. The effectiveness of the partial confinement is low and it is due to the failure of
the anchor bolts of the applied confinement. In figure 28b the observed behavior of specimen 3a Test 1 is also included. This is a virgin specimen that had a 5-CFRP layer partial confinement. Despite the increase in the CFRP layers, the observed effectiveness of the partial
confinement is low as was for specimen 1 test 2, again because of the failure of the anchor
bolts. In figure 29 the observed behavior of specimen 4 test 1 is depicted. This was a virgin
specimen with 5 CFRP layers partial confinement. In this case, a certain alteration was applied in the anchor bolts, by increasing their length. However, this was not sufficient to improve accordingly the effectiveness of the partial confinement, which was again linked to
the failure of the anchor bolts. The failed virgin specimen without the partial confinement is
shown in figure 30a whereas figure 30b depicts the failure of the anchor bolts for the "low
effectiveness" partial confinement.

3.2.2. Partial confinement of considerable effectiveness

In figures 31a and 31b the obtained behavior of specimen 5 (Test 3) with 5 (+2) CFRP layers is compared to specimen 1 test 1 (no partial confinement). Specimen 5 test 3 was formed by repairing a previously failed virgin specimen with special (low shrinkage) concrete as well as with 5 layers of CFRP forming the partial confinement. An additional two (+2) CFRP layers were applied at the part of the section where the anchor bolts were placed. This repaired specimen failed in compression with a modest increase (31%) in its capacity when compared with the capacity of the virgin unconfined specimen. The effectiveness of the partial confinement in this case was classified as considerable. This was due to an alteration in the anchoring of the partial confinement which proved to be relatively successful. The limit state for this specimen commenced with the tensile failure of the CFRP layers at the central zone and was accompanied, as expected, by a consequent compressive failure of the neighboring weak part of the section. This is depicted in figure 30c where the anchor bolts, which were left intact, are also shown. In figure 31b, the comparison is extended to include specimen 4 (Test 1), in which the partial confinement exhibited low effectiveness due to the failure of the anchor bolts.

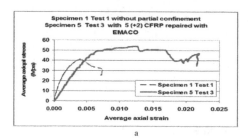
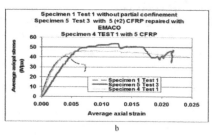

a   b

Figure 31. Partial confinement of considerable effectiveness. 31b. Partial confinement of considerable effectiveness.

3.2.3. Very effective partial confinement

In figure 32a, the behavior of specimens 5 (Test 4) and 5a (Test 2) is compared with the behavior of specimens 1 test 1 (with no partial confinement) and 5 (Test 3) discussed before. Specimens 5 (Test 4) and 5a (Test 2) were formed by repairing previously failed specimens with partial confinement of 7 layers of CFRP. Moreover, all the anchoring of their partial confinement was made with bolts going through the whole width of the strong part of the repaired section (see table 6 and figure 25). As can be seen in figure 32a, a substantial increase (40%), in the bearing capacity as well as in the deformability, resulted from the described partial confinement for these two specimens. Their behavior was in this way better than the behavior of specimen 4 (Test 1) which was classified before as one of considerable effectiveness of the partial confinement. Based in this increased capacity and deformability of specimens 5 (Test 4) and 5a (Test 2) they were classified as specimens with very effective partial confinement.

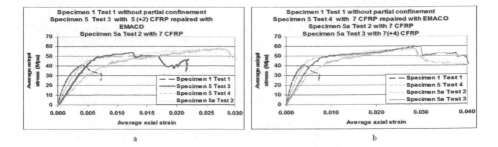

Figure 32. Very effective partial confinement. 32b. Very effective partial confinement.

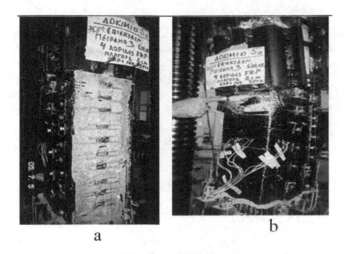

Figure 33. Strengthening stirrups - strong part. 33b. Failure of the CFRP layers of the mid-part.

The limit state of specimen 5 (Test 4) was accompanied with the failure of the stirrups of the strong part whereas the limit state of specimen 5a (Test 2) was accompanied with the tensile failure of the CFRP layers at the central zone. In figure 32b the observed behavior of specimen 5a (Test 3) is compared with that of specimens 5 (Test 4) and 5a (Test 2) previously discussed. Specimen 5a Test 3 had a partial confinement formed by 7 layers of CFRP. Moreover, the area sustaining the anchor bolts was strengthened by additional four (+4) layers of CFRP. Apart from the anchor bolts going through the whole width of the strong part of its section, the stirrups of this strong part were also strengthened (figure 33a). This time the increase in the bearing capacity was 46%; the increase in the deformability was also quite substantial. Thus, the behavior of this specimen is also one with very effective partial confinement. The observed limit state was accompanied with the compressive failure of the CFRP layers at the mid-part near the steel brackets (figure 33b).

3.3. Concluding remarks

a2. The undesired compression failure expected to develop in the base of vertical members with reinforced concrete cross sections having h/b ratio larger than 1.5 under combined vertical loads and seismic actions is studied through specially formed specimens subjected to uniform compression. The retrofitting of such specimens with partial CFRP confinement is aimed at prohibiting, up to a point, such compression failure. This type of partial confinement may also be applied to upgrading vertical structural members with non-accessible sides.

b2.. From the results of the experimental investigation with identical specimens, with or without this type of partial CFRP confinement, the successful application of such partial confinement was demonstrated. An increase of almost 50% was observed in the compression bearing capacity of some of the tested specimens. Moreover, the deformability of these specimens was substantially increased, demonstrating the effectiveness of this type of partial confinement.

c2. It was found from the experimental sequence that critical factors for this increase were the type of anchorage of the CFRP partial confinement and the number of CFRP layers. Successful anchoring of the CFRP layers allowed this partial confinement to become effective and to permit the use of a larger number of CFRP layers. In the present study alternative anchoring schemes were tried with limitations imposed by the geometry of the model cross-section. Similar limitations imposed by the geometry and the reinforcement of the cross-section will also dictate the design of such an anchoring scheme for a prototype cross-section. Further investigation on the performance of such prototype anchoring arrangements may be necessary.

4. Upgrading the flexural capacity of (R/C) vertical members

4.1. Flexural upgrading

The upgrading of the flexural capacity of reinforced concrete (R/C) vertical members with externally applied CFRP layers as means of tensile reinforcement is investigated here (see figure 34 and [4], [6], [11], [16], [22], [34], [40], [41])

The CFRP layers are applied at the two opposite sides of R/C specimens which were constructed for this purpose (figures 35a, 35b). They were designed with such longitudinal and transverse reinforcements shown in figure 35b that the flexural limit state would prevail. The specimens were tested to develop the ultimate flexural behavior with the corresponding damaged region concentrated near their joint with the foundation. These specimens had the foundation block fixed at the strong reaction frame of the laboratory of Strength of Materials and Structures of Aristotle University and were subjected to simultaneous constant vertical load as well as horizontal cyclic imposed displacements approximating thus the seismic actions. This loading sequence was applied prior to the application of any CFRP layers and was repeated again after these CFRP layers were attached to the damaged specimens. The observed upgrading of the flexural capacity in terms of ultimate overturning moment is presented and discussed.

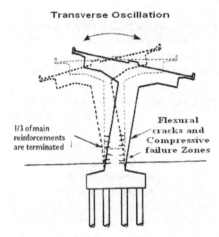

Figure 34. Flexural Damage

4.1.1. The virgin beam specimen

This specimen is shown in figure 35a with a rectangular cross-section of 200 mm x 500 mm and height h = 1815mm (from its top to the upper surface of its foundation). 8Φ6 was the longitudinal reinforcement that extended with no splices and was anchored to the foundation block and Φ6/100 closed stirrups the transverse reinforcement. The detailing of the cross-section is shown in figure 35b. The selected longitudinal and transverse reinforcement together with the loading arrangement, will cause the behavior of this virgin specimen to be dominated by the flexural rather than the shear mode of failure. More details are given in references [40] and [41].

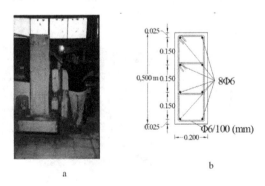

Figure 35. a Pier A. b. Cross-section and reinforcing details of Pier A.

4.1.2. The Loading sequence

The foundation of the specimen was anchored at the strong floor of a reaction frame. The specimen was then subjected to a simultaneous constant vertical load and a cyclic horizontal displacement with increasing amplitude in time, utilizing servo-electronically-controlled dynamic actuators. The frequency of this cyclic displacement was in some tests 0.1Hz whereas in other selected tests it became 1.0Hz. The vertical load was kept constant at 95KN. The horizontal imposed cyclic displacement was applied in 13 groups of continuously increasing amplitude. Each group included 3 full cycles of constant amplitude. This time history of the imposed displacement is depicted in figure 36a. The amplitude in this picture is non-dimensional and is given as a percentage of the final maximum amplitude of this cyclic loading. In the experimental sequence the maximum displacement horizontal amplitude was initially relatively small (2mm); as testing progressed, in subsequent cyclic loading sequences it reached values of 20mm to 25mm. The imposed displacements were measured at the location of the horizontal actuator which was placed at a height of 1400mm from the top of the foundation (figure 36b). Figures 36b and 36c depict this experimental set-up. As can be seen the specimen was placed in the strong reaction frame, having its foundation block, with dimensions in plan 1000mm by 1000mm and 300mm height, fully anchored to the strong floor. The horizontal and vertical actuators, as part of this strong reaction frame, applied the loading sequence described above. Instrumentation was provided in order to measure the variation of the applied horizontal and vertical loads as well as the most important aspects of the displacement field that resulted from the application of these loads to the specimen. The horizontal displacements of the specimens at the top were monitored as well as the displacements of the specimen at the region near the foundation block in order to identify the flexural, and shear deformations as well as the plastic hinge behavior at this part of the specimen. In figure 36b the two vertical sides of the specimen where the CFRP layers were attached are indicated together with the region where the anchoring arrangements were placed as part of the current investigation. Throughout these series of experiments the nominal thickness of the employed CFRP layers was 0.176mm with a given Young's Modulus E=350GPa. The measured maximum axial CFRP strain was approximately 1%. These CFRP layers were placed at these locations

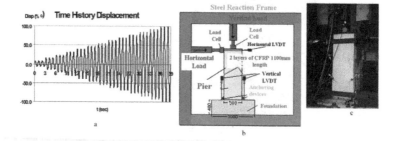

Figure 36. Imposed horizontal displacement. 36b. Experimental set-up. 36c Pier A placed at the strong reaction frame.

4.2. Observed Behavior - Results

4.2.1. The behavior of Pier A

The yield stress of the longitudinal reinforcements was equal to 344.8MPa whereas the concrete strength was equal to 21.2MPa. Based on these mechanical characteristics and the cross-section reinforcing detailing, predictions of the limit-state flexural behavior for this virgin specimen (without CFRP) were obtained in terms of the M-N interaction. The limit-state flexural behavior of a specimen with 2 layers of CFRP attached to both vertical sides of a repaired specimen is also obtained, based on the assumption that the cross-sections under flexure remain plane even with the attachment of the CFRP layers. The development of flexural cracks took place as shown in figure 37. At this horizontal cross-section the measured flexural behavior of Pier A is depicted in figure 38, which presents the variation of the resulting bending moment at this cross-section of the specimen against the rotation of the same cross section,. The rotation was obtained from measurements made by displacement transducers that were placed at the two sides of the pier monitoring the relative vertical displacement between the upper part of the specimen and its foundation block. Additional instrumentation was provided to monitor the rocking or the sliding of the foundation block itself, which proved to be non-significant. The observed maximum bending moment value compares quite well with the corresponding predictions of the limit-state flexure for this cross-section. The red line in these plots represents the corresponding envelope curve.

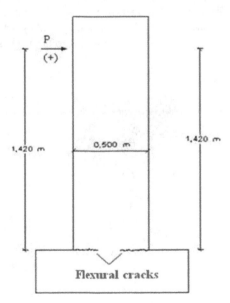

Figure 37. Observed flexural cracks for Pier A.

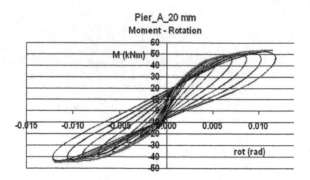

Figure 38. The measured flexural response of Pier A in terms of bending moment - plastic rotation.

4.2.2. Repaired specimen Rep-1 Pier A and observed behavior

The damaged Pier A was repaired in a way depicted by figures 39a,b. The resulting specimen is designated as Rep-1 Pier A. Two layers of CFRP were attached at each vertical side of the pier's cross-section. This attachment extended to a height of 1100mm from the foundation surface as well as at the top surface of the foundation block depicted in figure 39b. Moreover, in order to increase the bond strength between the CFRP layers and the top surface of the foundation block two double T steel sections were placed at these locations. A force normal to the bond surface at this location was applied by bolting these double T steel sections to the foundation block utilizing pre-stressing rods. Due to space limitations no further details are given here.

Figure 39. Specimen Rep-1 Pier A. Specimen 39b Rep-1 Pier A, attachment of CFRP layers. 39c Debonding of the CFRP layers from the right side of the repaired specimen Rep-1 Pier A.

4.2.3. Observed Behavior of Specimen Rep-1 Pier A

The loading sequence, described before, was applied to this repaired specimen. During increasing of the imposed horizontal displacement the debonding of the CFRP layer at the right side of the pier occurred as depicted in figure 39c. The obtained flexural behavior is depicted in figure 40 in terms of applied horizontal load against the imposed horizontal displacement at the top of the pier. If this behavior is compared with the corresponding behavior of the virgin specimen Pier A, a modest increase in its bearing capacity of approximately 20% can be observed as a result of the applied CFRP repair scheme, despite the debonding of the CFRP layers. The red line in this figure represents the corresponding envelope curve.

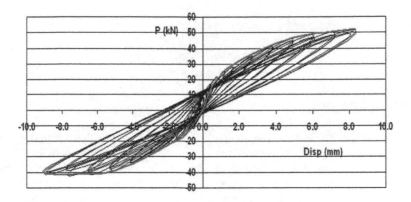

Figure 40. The measured flexural response of Rep-1 Pier A in terms of Horizontal load - Horizontal Displacement.

4.2.4. Repaired specimen Rep-2 Pier A and observed behavior

The damaged Rep-1 Pier A was repaired again in a way which is shown in figures 41a,b. The CFRP layers that were separated from the concrete at the right side of Rep-1 Pier A were reattached, both with the appropriate resin and additional 2mm diameter small bolts of 35mm length; these bolts were spaced at regular 50mm intervals along all the height of the CFRP layer at both sides. Moreover, the edge of the double T steel section neighboring the location of the CFRP layer at the pier-foundation joint was machined to form a curvature so that it provided a relatively uniform contact between the steel section and CFRP layers at this location. The locations of the bolts were first marked and drilled and then the bolts were applied including the corresponding plastic anchoring inserts. The specimen repaired in this way is designated as Rep-2 Pier A.

Figure 41. a The repair of specimen Rep-1 Pier A. b. The repair of specimen Rep-1 Pier A. c. Fracture of CFRP for specimen Rep-2 Pier A.

4.2.5. Observed Behavior of Specimen Rep-2 Pier A

The loading sequence, described in section 2, was also applied for this repaired specimen. During the increasing of the imposed horizontal displacement the fracture of the CFRP layers at the right side of the pier occurred as depicted in figure 41c. This fracture occurred at the location where the pier joins the foundation block. Figure 42a presents the variation of the measured horizontal load versus the horizontal displacement, whereas the resulting bending moment against the plastic hinge rotation is plotted in figure 42b. The plastic hinge which was formed during testing specimen Rep-1 was not repaired in any other way, but with the means described before. The fracture of the CFRP layer depicted in figure 41c took place at the same cross section. As was mentioned before the plastic rotation was measured by displacement transducers that were placed at the two sides of the pier. Because of the dislocation of one of these transducers the rotation angle after that was not monitored. This is noted at the bottom left hand side of figure 42b. Despite this lack of rotation measurements after the dislocation, the bending moment capacity remains approximately at the same levels, as deduced from the variation of the horizontal load (figures 42a and 43a). The effectiveness of the applied repair schemes for specimens Rep-1 and Rep-2 can be seen in figures 43a and 43b in terms of envelope curves. In figure 43a the comparison is made in terms of horizontal load – horizontal displacement whereas in figure 43b this is done in terms of bending moment-rotation. As can be seen, the effectiveness of the CFRP layers is inhibited by the debonding of the CFRP layers; thus although a considerable increase is achieved in the bearing capacity, this is not sustained in terms of displacement because of the premature debonding of the CFRP. The improvement of the bonding with the use of the employed bolting scheme for Rep-2 resulted in an increase of the bearing capacity both in terms of load and displacement and resulted in the fracture of the CFRP.

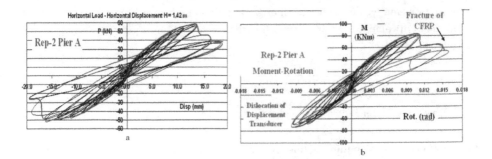

Figure 42. The measured flexural response of Rep-2 Pier A in terms of horizontal load - horizontal Displacement. 42b. The measured flexural response of Rep-2 Pier A in terms of bending moment - plastic rotation.

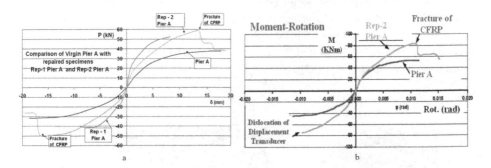

Figure 43. a Comparison of flexural response of repaired Rep-2 Pier A with repaired specimen Rep-1 Pier A and virgin specimen Pier A in terms of Horizontal load – Horizontal displacements. b. Comparison of flexural response of repaired Rep-2 Pier A with virgin specimen Pier A in terms of bending moment - plastic rotation.

From figure 43b the effectiveness of the repair scheme utilized in specimen Rep-2 Pier A can be deduced compared with the virgin specimen Pier A in terms of bending moment - plastic hinge rotation. The achieved increase in bending moment is almost 50% and the flexural behavior in terms of rotation is satisfactory up to the fracture of the CFRP layer at the joint between the pier and the foundation. Comparing the maximum value of the bending moment (83KNm) that was resisted by specimen Rep-2 Pier A with the one predicted (130KNm), as an ultimate bending moment that the cross-section reinforced by the applied CFRP layers can ideally resist, it can be concluded that the fracture of the CFRP did not allow this ideally maximum bending moment value to be reached. One reason for this is the way the anchoring of the CFRP layers at the foundation block is achieved with the utilization of the double T steel sections described before. Despite the machining of the edge of these double T steel sections the development of high stress levels concentrated at the CFRP layers at these locations is the cause of the observed CFRP fracture. Moreover, such an anchoring scheme is quite impractical for prototype conditions. For these reasons a different anchoring scheme was investigated in order to address both these shortcomings that were found to be critical in the repair schemes investigated so far. This is presented next in a summary form.

4.2.6. Repaired specimen Rep-3 Pier A and observed behavior

Figure 44a depicts the outline of this anchoring scheme and figure 44b depicts the application of this anchoring scheme for repair specimen Rep-3 Pier A, whereby the CFRP layers are folded around a cylinder that is anchored to the foundation [40]. Moreover, in order to avoid the separation (detachment) of the CFRP layers from the sides of the specimen, as observed in specimen Rep-1, the bolting scheme that was used in specimen Rep-3 was further improved by using bolts and washers of larger size (due to space limitations no details are given here). In addition, to avoid the development of shear failure at the bottom part of the pier, three horizontal CFRP closed hoops were attached in this region of the specimen (see figures 44c, 44d and [40], [41], [42]).

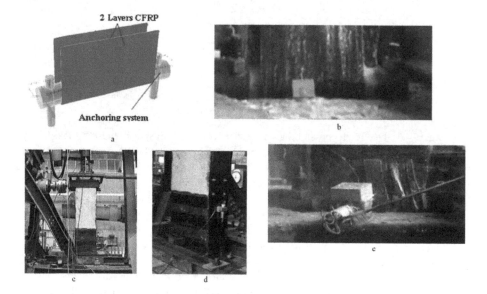

Figure 44. a New anchoring device. b. New anchoring device. c. Specimen Rep-3 Pier A. Figure d. Specimen Rep-3 Pier A. Figure e. Partial fracture of CFRP layers at the cylindrical part of anchoring device

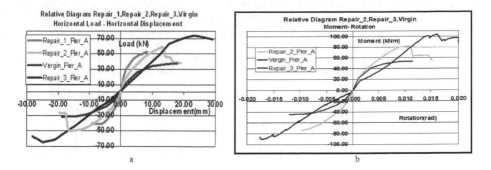

Figure 45. Comparison of flexural response of repaired specimens Rep-3, Rep-2, Rep-1 and virgin specimen. 45b. Comparison of flexural response of repaired Rep-3, Rep-2 and virgin specimens.

4.2.7. Observed Behavior of Specimen Rep-3 Pier A

The loading sequence, described before, was also applied for this repaired specimen. During increasing of the imposed horizontal displacement the partial fracture of the CFRP layer at the right side of the pier occurred as depicted in figure 44e. This fracture occurred at the location where the CFRP layers are folded around the cylindrical part of the anchoring device of figures 44a,b. In figures 45a and 45b the measured flexural behavior in terms of horizontal load-horizontal displacement and bending moment-rotation envelope curve diagrams for repaired specimen Rep-1, Rep-2 and Rep-3 is compared with the coresponding flexural behavior of the virgin specimen Pier A. Due to space limitations the measured full cyclic response for specimen Rep-3 is not included in this presentation. The following observations can be made on the basis of these diagrams. a) The horizontal load and maximum bending moment bearing capacity of the latest repair scheme, namely specimen Rep-3, exhibits an increase of almost 100% when compared with the corresponding bearing capacity of the virgin specimen. b) When this comparison is made between the latest repair scheme Rep-3 and re pair scheme Rep-2 the increase is of the order of 25%. Moreover, the new anchoring scheme does not cause an abrupt fracture of the CFRP at the anchoring device. This is reflected in figures 45a and 45b by the fact that the obtained response of specimen Rep-3 extends to relatively large displacement and rotation values. c) A reduction in the initial stiffness appears to be present in specimen Rep-3 when its obtained cyclic response in terms of the envelope curve of figures 45a,b is compared to the corresponding curves of specimen Rep-2 and the virgin specimen. This must be attributed to the fact that most parts of specimen Rep-3 have been subjected to three full loading sequences except the new parts of the CFRP. d) Rep-1 corresponds to the least effective repair scheme due to the debonding of the CFRP layers.

4.3. Concluding Observations

a3. The upgrading of the flexural behavior of vertical R/C structural elements was investigated utilizing the possibility of attaching longitudinal CFRP layers that can be stretched developing tensile forces externally at the opposite sides of these elements. Critical aspects for their satisfactory performance are the bonding of the CFRP layers as well as the effective transfer of the tensile forces to the foundation block.

b3. The bonding of the CFRP layers at the sides can be improved by bolting schemes that, in this way, increase the cooperation between the CFRP layers and the concrete part.

c3. The transfer of the tensile forces that develop at the CFRP layers to the foundation is a difficult technical problem. Two schemes were tried. The 2nd scheme is more practical and was successful in mobilizing sufficiently the available capacity of the applied CFRP layers in such a way that it resulted in approximately 100% increase in the flexural capacity when compared to the initial flexural capacity of the virgin specimen before the application of the CFRP layers.

d3. The applicability of such an anchoring system must be validated in a more general way before final practical conclusions can be reached.

e3. Such an upgrading of the flexural capacity may lead to the appearance of the shear mode of failure. However, the shear capacity can be increased with relative ease by closed CFRP hoops.

5. Relevant code provisions - Emphasis in the application of FRP strips for shear strengthening

During the last fifteen years numerous researchers ([12], [18], [29], [34]) have proposed different approaches for predicting the behavior of FRPs when they are used for strengthening structural members. Several design guidelines and code provisions adopted some of the proposed approaches. More specifically, the approach of Triantafillou and Khalifa [33] has been adopted by the American Code ACI 440 [11] whereas the model of Chen and Teng ([18], [25]) is utilized in the Greek Code of Interventions [6]. Eurocode 8 part 3 ([7]) follows the guidelines of FIB [8]. In the following paragraphs the aforementioned three Codes are discussed with particular emphasis being given to the application of FRP strips for shear strengthening. All codes recommend a limit-state tensile force that a cross-section (A_{FRP}) of an FRP sheet can sustain. This limit-state tensile force is obtained through equation 2:

$$V_{FRP} = \frac{A_{FRP} \cdot \varepsilon_{FRP} \cdot E_{FRP}}{\varphi} \tag{2}$$

Where: V_{FRP}: Limit-state tensile force of the FRP strip, φ: safety factor, A_{FRP}: cross section area of a strip of FRP

ε_{FRP}: developed axial strains on the FRP strip, E_{FRP}: modulus of elasticity of the FRP strip

The various codes recommend the application of Equation 2 by defining various values for the safety factor (φ) and the limit-state axial strain value (ε_{FRP}). These values vary according to the application depending on whether the predicted FRP contribution is for shear, flexure or confinement. In doing so, the basic factors that are varied among the Codes are the safety factor and the allowable developed strain on FRP. Table 7 presents the maximum allowable ε_{FRP} values and the recommended safety factor (φ) values according to ACI 440, Eurocode part 3, and the Greek Code for structural interventions, when anchoring is used in the transfer of FRP strip forces.

Code	Max allowable ε_{FRP}		φ
	Flexure	Shear	
ACI 440 [11]	$0.9\,\varepsilon^{*}_{nom}$	0.4 %	1.24 - 1.47
Eurocode 8 (part3) 7	-	0.6 %	1.5
Greek Code of Interventions 6	$\varepsilon^{*}_{nom}\,/\,2 < 1.5\%$	$\varepsilon^{*}_{nom}\,/\,2 < 1.5\%$	1.25

Table 7. Maximum allowable developed strains ε_{FRP} and proposed safety factors φ when anchoring is used. Where: ε^{*}_{nom}: max strain from manufacturer

The allowable limit-state axial strains (ε_{FRP}) value defined by the various codes also depend upon the way the FRP sheets are attached upon the structural elements in need of strengthening. According to all code provisions, there are two basic ways of attaching FRP sheets on structural elements. First, they can be attached using an anchoring device or they can be fully wrapped around the structural element. Secondly, they can simply be attached on the surface of the structural element through an organic or inorganic matrix in an open loop U-shaped without any anchoring. When the simple attachment is used without anchoring the debonding mode of failure prevails, whereas when FRP sheets are combined with an anchoring device or when they are fully wrapped around a column or a beam, fracture of FRP's occurs. The debonding mode of failure poses a limitation to the FRP axial strains and stresses. This limitation is reflected in all three Codes through a factor (usually designated as k_{v}), that takes values less than 1. This reduction factor depends on the recommended value of the attachment FRP length (effective attachment length). Table 8 lists the basic formulae included in the various codes for calculating this effective attachment length. The most important factor in calculating the effective attachment length is the determination of bond strength, which is not easily obtained since it depends on the actual concrete tensile strength and the state of the surface where the FRP is attached; both these parameters can easily vary even for the same structural element.

When the attachment of FRP sheets is combined with an anchoring device a more reliable transfer of forces can be achieved thus resulting in an equally reliable design of the relevant strengthening scheme. Moreover, since the prevailing mode of failure is the fracture of the FRP's, when anchoring of the FRP strips is used, the exploitation of the material of the FRP sheets is enhanced by reaching higher values of axial strains than for the cases governed by the debonding mode of failure. Under all circumstances the maximum allowable strains should not exceed the value presented in table 7. All codes demand the use of a safe anchoring device that would allow the fracture of FRP sheets without proposing any calculations for the safe design of these anchoring devices. In any case, equation (3) describes this condition for the bearing capacity of a permissible anchoring device

Code	Effective Length (Le)	
ACI 440 [11]	$L_e = \dfrac{23300}{(nt_f E_f)^{0.58}}$	where n: number of FRP layers t_f: thickness of FRP E_f: Modulus of elasticity
Eurocode 8 (part3) [7]	$L_e = \sqrt{\dfrac{E_f \cdot t_f}{\sqrt{4 \cdot \tau_{max}}}}$	where E_f: Modulus of elasticity t_f: thickness of FRP τ_{max}: bond strength
Greek Code of Interventions [6]	$L_e = \sqrt{\dfrac{E_j t_j}{2 f_{ctm}}}$	where E_j: Modulus of elasticity t_j: thickness of FRP f_{ctm}: tensile concrete strength

Table 8. Calculation of effective length. Where V_{FRP}: total force received by FRP, $V_{anchoring\ device}$: total strength of the anchoring device.

$$V_{FRP} \leq V_{anchoring\ device} \tag{3}$$

6. Special study for anchoring FRP strips

In this section results from a recent research effort conducted at the Laboratory of Strength of Materials and Structures of Aristotle University will be briefly presented and discussed ([13], [42]). This research aimed to investigate the effectiveness of a specific FRP strip anchoring device by utilising a number of small concrete prismatic specimens, that can house such an FRP strip with sufficient width and length. In all, twelve (12) specimens were investigated. For six specimens no surface preparation of the concrete specimen was applied when attaching the FRP layers, whereas six other specimens had their surface treated according to construction guidelines. CFRP layers were attached to all specimens ([28], [43]). The use of the anchoring device was utilised on three (3) specimens with treatment preparation and on three specimens without such treatment. All the concrete prisms were fabricated using the same concrete mix and the same internal reinforcement, which was used to prohibit any accidental failure. The measured cylinder strength of the concrete was equal to 22 MPa. The properties of the used CFRP are listed in Table 9, as given by the manufactures.

Material	Type / Name	Modulus of Elasticity (GPa)	Thickness of Layer (mm)	Ultimate strain
CFRP	SikaWrap 230 C/45	234	0.131	0.018

Table 9. Properties of the used FRP sheets.

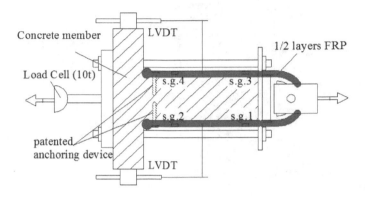

Figure 46. Experimental Set-up.

The loading arrangement is depicted in figure 46, whereby the tensile force is directly applied in the axis of symmetry at the right part of the FRP strips that forms an open hoop at this location; the other two sides of the FRP strip are bonded in a symmetric way on the top and bottom side of the concrete prism, as shown in this figure. When anchors were employed they were added at these locations (ends of the FRP strips). Despite the symmetry of this test set-up, instrumentation was provided that was able to record symmetric as well as asymmetric response of the specimen, especially during the initiation and propagation of the debonding process. During testing, the applied load is measured together with the longitudinal (axial) strains at four different locations of the external surface of the FRP strip, as indicated in figure 46 (s.g.1 to s.g.4), in order to calculate the stress field that develops at the FRP layer before and during the debonding.

Moreover, the relative longitudinal displacement between the concrete prism and the FRP surface is also monitored using four displacement transducers that are properly attached to the specimen, as indicated in this figure, in order to record the initiation and propagation of the debonding of the FRP. The used anchoring device was developed at the Laboratory of Strength of Materials and Structures of Aristotle University of Thessaloniki in Greece and it is patented with patent number WO2011073696 [42]. Figure 47 presents some details of this device. The tested specimens with their details are listed in table 10 together with their code names. The first letter C in the code name denotes a carbon fiber reinforcing polymer strip. (CFRP). The type of surface preparation is denoted by the second letter of the code name (S for smooth surface, R for rough surface; the type of anchor is denoted by the third letter of the code name (N for no anchor and P for patented anchoring device). Moreover, the number of layers of these strips is denoted by the fourth character of the code name (1 for one layer and 2 for two layers). The tests were conducted using a 1000 kN capacity hydraulic piston. The measurements of load, displacements and strains were recorded using an automatic data acquisition system.

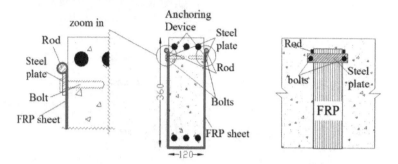

Figure 47. Patented Anchoring Device (WO2011073696).

Spec. Name	Material Type	Number of Layers	Surface Type	Anchor Type	No. of specimens	Bolt Type
CSN1	CFRP	1	smooth	no	3	no
CRN1	CFRP	1	rough	no	3	no
CSP2	CFRP	2	smooth	patented	3	2XHUS by Hilti
CRP2	CFRP	2	rough	patented	3	2XHUS by Hilti

Table 10. Details of Specimens.

Spec. Name	Anchor Type	Max Load (kN)	Failure Mechanism	Load at Debonding (kN)	Max Strain (μStrain) s.g.1	Max Strain (μStrain) s.g. 3	Material Exploitation M_e	Load from Strain (kN)
CSN1	no	27.9	debonding	27.9	5400	5400	0.30*	32.5
CRN1	no	42.7	debonding	42.7	6335	7215	0.38*	40.8
CSP2h	patented	112.8	CFRP fracture	30	9810	8320	0.50	109.2
CRP2h	patented	103.0	CFRP fracture	40	7220	9330	0.46	99.7

Table 11. Summary of experimental results. * Specimens with only one CFRP layer. If for these specimens a second CFRP layer was added without anchor, due to the debonding failure, no increase in the maximum load and maximum strain can be achieved. Consequently, in such a case the exploitation ratio value would be half the values listed in this table.

Experimental Results and Discussion: The summary of the experimental results is listed in table 11. In this table, the observed failure mechanism is listed together with the corresponding value of the ultimate measured load as well as the value of the load recorded at the initiation of debonding. Moreover, the maximum strain values measured by the strain gauges at locations 1 and 3 on the FRP strip surface are also listed. The average value of these maximum FRP strains was utilised to calculate indirectly the load sustained by the FRP strips taking into account their total cross-sectional area and the value of the Young's modulus, listed in table 10. Finally, the material exploitation indicator (M_e) is given in the same table as the ratio of the maximum measured strain by the ultimate strain value provided by the manufacturer (Table 10).

For specimens CSN1 having non-prepared surfaces, an average ultimate load was found equal to 27.9 kN. The equivalent ultimate load value when prepared contact surfaces were used, specimens CRN1 and SRN1, was 41 kN. Thus, the proper preparation of the contact surface resulted in a 46% increase of the ultimate load. This increase that is attributed to the surface preparation was observed when no anchoring device was utilized. The employed surface preparation needs at least twice as much time as when no surface preparation is made.

When an anchoring device is employed, it can be seen that preparation of the concrete contact surface is of no significance (see the ultimate load values of the anchored specimens of table 11 and those of debonding). In all tested cases the ultimate load is greater than the load at debonding. Thus, when using an effective anchoring device, the cost of properly treating the contact surface can be avoided.

The load at debonding for those specimens whose surfaces were not specially treated had a value approximately equal to 30 kN with small deviations. Similarly, the load at debonding for those specimens whose surfaces were specially treated had a value approximately equal to 40kN with small deviations. When an anchoring device is utilized the ultimate capacity increases from 28 kN to 112.8 kN for the set of specimens strengthened with CFRP. This represents a fourfold increase in the value of ultimate load.

Finally, when the patented anchoring device was applied for specimens CSP2h, CRP2h, a significant increase in the bearing capacity was observed. This time the performance of the anchoring device was very satisfactory and the observed failure was that of the fracture of the FRP strips for all these specimens. Figure 6 depicts such a failure mode for specimen SSP2b (see also table 11).

In order to discuss the observed behavior in terms of exploitation of the high strength of the FRP materials the following procedure was used. As already mentioned, a material exploitation indicator was found (M_e) as the ratio of the maximum measured strain values (average of the two sides, Table 3) for each specimen over the ultimate strain values as they are measured for the used FRP materials (see table 9). These material exploitation indicator values are also listed in Table 11, having ideally as an upper limit the value of 1. As can be seen in this table, the highest M_e value during the present experimental sequence reaches the value 1 and this was achieved by the specimen that utilises the patented anchoring device together with two layers of CFRP strips. As expected, debonding of the FRP strips or failure of the anchoring device results in relatively low values of the material exploitation indicator M_e. (0,60 and 0,79) This research effort is in progress experimenting with various alternative anchoring details.

Figure 48. Detail of the anchoring device and the fracture of the CFRP strip when such an anchoring device is employed.

6.1. Concluding Observations

a4. Proper preparation of the contact surface between the FRP strips and the concrete face resulted in a 46% increase of the ultimate load. This increase that is attributed to the surface preparation was observed when no anchoring device was utilized. The employed surface preparation needs at least twice as much time as when no surface preparation is done.

b4. When an anchoring device is employed, it can be seen that preparation of the concrete contact surface is of no significance. With the proper anchoring device the ultimate capacity increases four times.

c4. It is important to properly detail the anchoring device in order to drive the mode of failure to the fracture of the FRP strip rather than the failure of the anchor thus exploiting the high tensile strength FRP potential.

d4. The highest value of the FRP material exploitation indicator was achieved in the specimen that utilises the anchoring device patented by Aristotle University together with two layers of CFRP strips.

e4. As expected, debonding of the FRP strips or failure of the anchoring device results in relatively low values of the material exploitation indicator M_e. This research effort is in progress experimenting with various alternative anchoring details.

7. Basic qualification tests for fiber reinforcing polymers (FRP) sheets to be used in dealing with earthquake structural damage.

In what follows, a brief description is given of basic qualification tests to be performed with FRP sheets used in repair / strengthening schemes of R/C structural elements in the framework of earthquake structural damage. As already discussed in sections 5 and 6, the main critical parameters for such use of FRP sheets are: a) the modulus of elasticity, b) the maximum axial strain and c) the effective thickness of these sheets. These properties are usually provided as technical specification data by the manufacturers of these materials. Moreover, the manufacturers of the FRP sheets for this type of application also provide technical information on the organic or inorganic matrices that are compatible with the relevant FRP sheets and the type of material surface of the structural element to which these sheets are to be externally applied. Finally, the technical information of the manufacturers as well as the relevant code guidelines include preparatory actions which must be taken before the FRP sheets are applied, such as concrete surface preparation, rounding of corners and proper application of the matrices together with the FRP sheets prohibiting the formation of any air pockets. In case the tensile characteristics of the FRP sheets must be confirmed, specimens of the FRP material must be taken in order to define the modulus of elasticity (E), the Poisson's ratio (v), the tensile strength and the maximum tensile failure strain (ε_m). These tests are performed following the appropriate specifications of the relevant standard (e.g. European Standard EN ISO 527-5: 1997).

During the various experimental sequences that were presented in the preceding sections 2, 3 and 4 the employed FRP sheets were tested in order to verify their basic tensile characteristics. Selective measured values are reported in the relevant sections in brief (see subsections 2.2.2, 3.1.2 and 4.1.2). For all tests reported in sections 2, 3 and 4 the FRP sheets were attached to the reinforced concrete specimens using the same organic matrix. This was a two part, solvent free, thixotropic epoxy based impregnating resin / adhesive. The density of this epoxy resin is 1.31 kg/l and its average viscosity is approximately 7.000 mPas. The Thermal Expansion Coefficient is equal to 45 x 10-6 per °C, its tensile strength according to DIN 53455 is 30 Mpa, the Young's Modulus is 4500 Mpa and the ultimate elongation is 0.9%. There is a manufacturers warning that this product is not suitable for chemical exposure.

7.1. Bond Strength between CFRP layers and Concrete Substrate.

Another property of interest is the bond strength between the FRP sheets and its matrix with the surface of the structural member, especially when the FRP sheets are not accompanied by the appropriate anchoring, as discussed in the preceding sections. There are certain bond strength tests aiming to insure that the debonding mode of failure does not occur between the matrix and the FRP sheet (e.g. European Standard EN ISO 1542: 1999). However, when designing for the debonding limit-state use is made of the tensile strength of the concrete substrate instead of this bond strength. A special investigation was performed at the Laboratory of Strength of Materials and Structures of Aristotle University aiming at measuring directly this bond strength of the employed CFRP sheets attached with the named above epoxy resine to the concrete substrate. Figure 49a depicts this simple test aimed at measuring the bond strength between the employed CFRP layers and the concrete surface (ref. [40], [41]). Figure 49b depicts the obtained results together with a best-fit linear variation of the bond strength (τ) versus the applied normal stress ($\sigma_{\varepsilon\gamma\kappa}$). As expected, a modest increase can be obtained in the bond strength (τ) between the CFRP layers and the concrete surface if a normal stress ($\sigma_{\varepsilon\gamma\kappa}$) increase is applied at the bond surface. This was partly made use of in the various anchoring schemes employed in sections 2, 3 and 4.

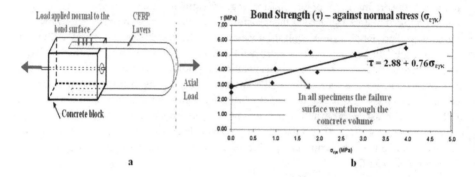

Figure 49. a Test for estimating the bond between CFRP layers and concrete surface and b obtained results.

7.2. Cost estimates

From a comparative study of data dealing with typical cases of repair and strengthening of R/C structural elements with conventional methods (jacketing with gunite or cast-in-place R/C concrete) versus CFRP based strengthening schemes the following constitute the average findings. The cost of a fully wrapped CFRP strengthening scheme when compared to a cast-in-place jacketing, represents an increase of approximately 20% to 30%. However, when a jacketing strengthening scheme with gunite (shotcrete) is applied instead of FRP's the cost increase is of the order of 30% to 40%. If the CFRP repair / strengthening solution does not include full wrapping the above cost increase is expected to be reduced.

8. Conclusions

1. This presentation dealt with repair and strengthening schemes of earthquake damaged reinforced concrete (R/C) structural elements utilizing externally attached fiber reinforcing plastics (FRP's). Such strengthening schemes were studied when applied to slabs, beams and vertical structural members. The success of such an upgrading scheme was discussed together with its limitations on the basis of a series of relevant experimental results.

2. One of the main limitations results from the way the tensile forces which develop on these FRP sheets can be transferred. When the transfer of these forces relies solely on the interface between the FRP sheet and the external surface of the reinforced concrete structural elements, the delaminating (debonding) mode of failure of these sheets occurs, due to the relatively low value of either the ultimate bond stress at this interface or

the relatively low value of the tensile strength of the underlying concrete volume. This mode of failure is quite common and it occurs in many applications well before the corresponding FRP sheets develop tensile axial strains in the neighborhood of values mentioned before as design limit axial strains (approximately of the order of 1%).

3. Alternative ways of transferring these tensile forces, apart from the simple attachment, in order to enhance the exploitation of the FRP material potential have been also presented and discussed based on experimental evidence from ongoing research at Aristotle University.

Acknowledgements

Partial financial support for this investigation was provided by the Hellenic Earthquake Planning and Protection Organization (EPPO), which is gratefully acknowledged.

Author details

George C. Manos* and Kostas V. Katakalos

*Address all correspondence to: gcmanos@civil.auth.gr

Laboratory of Experimental Strength of Materials and Structures, Department of Civil Engineering, Aristotle University of Thessaloniki, Greece

References

[1] Papazachos, B. C. (1990). Seismicity of the Aegean and the Surrounding Area. *Tectonophisics*, 178, 287-308.

[2] Manos, G.C. (2011). Consequences on the urban environment in Greece related to the recent intense earthquake activity. Int. Journal of Civil Eng. and Architecture Dec Serial No. 49) , 5(12), 1065-1090.

[3] Paz, M. (1994). International Handbook of Earthquake Engineering: "Codes, Programs and Examples". edited by Mario Paz, Chapter 17. , Greece by G.C. Manos, Chapman and Hall 0-41298-211-0

[4] Pauley, T., & Priestley, M. J. N. (1992). Seismic Design of Reinforced Concrete and Masonry Structures. J. Wiley & Sons, INC USA.

[5] Organization of Earthquake Planning and Protection of Greece (OASP),. (2001). *Guidelines for Level- A earthquake performance checking of buildings of public occupancy»*, *Athens*.

[6] Organization of Earthquake Planning and Protection of Greece (OASP), (2011). Guidelines for Retrofitting in Reinforced Concrete Buildings, Athens.

[7] Eurocode-8. (2004). Design of structures for earthquake resistance- Part 2: Bridges, DRAFT. 2004, *European Committee for Standardization* [3].

[8] F.fF. I. B. I.S. C. (2001). Externally Bonded FRP Reinforcement for RC Structures. " in Bulletin 14, f.T.G. 9.3, Editor. , 165.

[9] Greek Code for the Design of Reinforced Concrete Structures" (2000). in Greek).

[10] Provisions of Greek Seismic Code. (2000). *OASP, Athens, December 1999. Revisions of seismic zonation introduced in 2003.*

[11] ACI, (2008). Guide for the Design and Construction of Externally Bonded FRP Systems for Strengthening Concrete Structures (ACI 440.2R-08). in ACI 440.2R-08 American Concrete Institute: Farmington Hills , 45.

[12] Bakis, C., et al. (2002, May). Fiber-Reinforced Polymer Composites for Construction-State of the Art Review'. *Journal of Composites of Construction, ASCE.*

[13] Katakalos, K., Manos, G. C., & Papakonstantinou, C. G. (2012). Comparison between Carbon and Steel Fiber Reinforrced Polymers with or without Ancorage. ", 6th International Conference on FRP Composites in Civil Engineering Rome

[14] Manos, G. C., Kourtides, V., & Matsukas, P. (2007). Investigation of the flexural and shear capacity of simple R/C beam specimens including repair schemes with fiber reinforced plastics. FRPRCS-8 World Conference, Greece, July 16-18 2007

[15] Nanni, A. (1995). Concrete repair with externally bonded FRP reinforcement. *Concrete International*, 17(6), 22-26.

[16] Teng, J. G., Chen, J. F., Smith, S. T., & Lam, L. (2002). FRP strengthened RC structures. Chichester: John Wiley and Sons.

[17] Papakonstantinou, C. G., & Katakalos, K. . (2009). . Flexural Behavior of Reinforced Concrete Beams strengthened with a hybrid retrofit system. Structural Engineering and Mechanics Techno Press 31-5

[18] Chen, J. F., & Teng, J. G. (2003). Shear Capacity of Fiber-Reinforced Polymer-Strengthened Reinforced Concrete Beams: Fiber Reinforced Polymer Rupture. Journal of Structural Engineering May 1 ASCE, 5 , 129(5), 615-625.

[19] Papakonstantinou, C. G., Katakalos, K., & Manos, G. C. (2012). Reinforced Concrete T-Beams Strengthened in Shear with Steel Fiber Reinforced Polymers. Paper presented at 6th International Conference on FRP Composites in Civil Engineering, Rome.

[20] Chen, J. F., & Teng, J. G. (2001). Anchorage strength models for FRP and steel plates bonded to concrete. *Journal of Structural Engineering, ASCE*, 127(7), 784-791.

[21] Lamanna, A. J., Bank, L. C., & Scott, D. W. (2004, May/June). Flexural strengthening of R/C beams by mechanically Attaching FRP strips. *Journal of Composites of Construction, ASCE*, 203-210.

[22] Lee, J. H., Lopez, M. M., & Bakis, C. E. (2007, July). Flexural Behavior of reinforced concrete beams strengthened with mechanically fastened FRP strip. Paper presented at FRPRCS-8, World Conference,, University of Patras.

[23] Paterson, J., & Mitchell, D. (2003, May). Seismic Retrofit of Shear Walls with Headed Bars and Carbon Fiber Wrap. *Journal of Structural Engineering, ASCE*, 606-614.

[24] Manos, G. C., & Papanaoum, E. (2009, June). Earthquake Behavior of a R/C Building Constructed in 1933 before and after its Repair. *STREMAH, Tallin*, 22-24.

[25] Chen, J. F., & Teng, J. G. (2003). Shear capacity of FRP strengthened RC beams: FRP debonding. *Construct Build Mater*, 17(1), 27-41.

[26] Ekenel, M., Rizzo, A., Myers, J. J., & Nanni, A. ". (2006, September/October). Flexural Fatigue Behavior of Reinforced Concrete Beams Strengthened with FRP Fabric and Precured Laminate Systems. *Journal of Composites of Construction, ASCE*, 433-442.

[27] Khalifa, A., & Nanni, A. (2002, April). Rehabilitation of rectangular simply supported RC beams with shear deficiencies using CFRP composites. *Construction and Building Materials*, 16(3), 135-146.

[28] Manos, G. C., Katakalos, K., & Papakonstantinou, C. G. (2011). Shear behavior of rectangular beams strengthened with either carbon or steel fiber reinforced polymers. *Applied Mechanics and Materials (Trans Tech Publications)*, 82, 571-576.

[29] Triantafillou, T. C., & Antonopoulos, C. P. (2000). Design of concrete flexural members strengthened in shear with FRP. Journal of Composites for Construction November ASCE., 4(4)

[30] Kani, G. N. J. (1966, Jun). Basic Facts Concerning Shear Failure. *J. ACI*, 63, 675-692.

[31] Zararis, P. (2002). Reinforced Concrete. Kyriakides Editions, Thessaloniki (in Greek).

[32] Zsutty, T. C. (1968, November). Beam Shear Strength Prediction by Analysis of Existing Data. *Journal ACI*, 65, 943-951.

[33] Triantafillou, A. (2003). Software for the design of reinforced concrete elements strengthened with fibre reinforced polymers FRP. (in Greek).

[34] Pantazopoulou, S. J., Bonacci, J. F., Sheikh, S., Thomas, M. D. A., & Hearn, N. (2001). Repair of corrosion-damaged columns with FRP wraps. *Journal of Composites for Construction*, 5(1), 3-11.

[35] Tastani, S. P., Pantazopoulou, S. J., et al. (2006). Limitations of FRP Jacketing in Confining Old-Type Reinforced Concrete Members in Axial Compression. Journal of Composites for Construction February 1,. ASCE., 10(1)

[36] Manos, G. C., & Kourtides, V. (2007, July 16-18). Retrofitting of long rectangular R/C Cross-Sections with Partial Confinement employing Carbon Fiber Reinforcing Plastics. Paper presented at FRPRCS-8 Conf., Greece. (128).

[37] Tsonis, G. (2004). Seismic Assessment and Retrofit of Existing Reinforced Concrete Bridges. Ph.D. Thesis, Politecnico di Milano.

[38] Kawashima, K. (2000). Seismic performance of RC bridge piers in Japan: an evaluation after 1995 the Hyogo-ken nanbu earthquake. *Prog. Struct. Engng Mater*, 2, 82-91.

[39] Pinto, A. V. (1996, November). Pseudodynamic and Shaking Table Tests on R.C. Bridges. *ECOEST PREC*8 R* [8].

[40] Manos, G. C., Katakalos, K., Kourtides, V., & Mitsarakis, C. (2007, July). Upgrading the Flexural Capacity of a Vertical R/C Member Using Carbon Reinforcing Plastics Applied Externally and Anchored at the Foundation. Paper presented at FRPRCS-8 World Conference.

[41] Manos, G. C., Katakalos, K., & Kourtides, V. (2008). Study of the anchorage of Carbon Fiber Plastics (CFRP) utilized to upgrade the flexural capacity of Vertical R/C members. Paper presented at 14WCEE, Beijing, CHINA.

[42] Manos, G.C., Katakalos, K., & Kourtides, V. (2011). Construction structure with strengthening device and method. European Patent Office, Patent A1) —2011-06-23(WO2011073696)

[43] Manos, G.C., Katakalos, K., & Kourtides, V. (2011). The influence of concrete surface preparation when fiber reinforced polymers with different anchoring devices are being applied for strengthening R/C structural members. *Applied Mechanics and Materials*, 82, 600-605.

Chapter 5

Hybrid FRP Sheet – PP Fiber Rope Strengthening of Concrete Members

Theodoros C. Rousakis

Additional information is available at the end of the chapter

1. Introduction

Fiber Reinforced Polymer (FRP) reinforcements are extensively used in the strengthening of existing concrete members. FRP sheets consisting of epoxy resins and carbon, glass, aramide etc. fibers can serve reliably as flexural, shear or confinement reinforcement. Adequate external confinement of concrete by elastic materials leads to far higher effectiveness in strength and strain enhancement of concrete under compression than common steel, since the steel yields. However the use of impregnating resin in FRPs results in a composite material that resists relatively low working temperature. It also requires suitable environmental conditions during impregnation

Significant recent research efforts focus on the overcoming of such disadvantages relative to the use of polymers. The substitution of the organic resins by inorganic cement based binders seems a viable option ([1] among else). The inorganic mortar serves as a matrix that interacts with the grid reinforcement that consists of textiles. Textile reinforced mortars (TRM) have been already used as confining or shear reinforcement successfully.

A few investigations deal with rope reinforcements made of aramide or vinylon fibers. Those ropes are used as external strengthening or internal shear reinforcement. They combine easy handling and low sensitivity to local damage of fibers due to bending or small corner radius or scratching or stress concentrations [2]. Furthermore, there is no need for impregnating resins or binders especially in external confinement applications of ropes. Polypropylene is an ultra high deformability material, recognized as an effective mass reinforcement in concrete. Peled [3] uses polypropylene tubes to confine concrete. However the provided lateral confinement was low and the effects on concrete were limited. Numerous researches look into the response of concrete columns wrapped by carbon FRP sheets of high modulus of elasticity [4, 5, 6, 7 & 8 among else]. FRP confined concrete may present a considerable strength and strain enhancement. However the efficiency of the wraps is limited by the low deformability of the FRP. Confinement with materials of different modulus of elasticity and of the same lateral rigidity (E_l) will provide higher axial strain ductility to concrete according to their deformability [9]. Thus, materials of very low elastic modulus which have high strain at failure may provide considerable confinement of concrete.

Considering the advantages of composite ropes for structural applications, an experimental research program is presented that investigates the use of fiber ropes as external confining reinforcement on standard concrete cylinders already confined by glass FRP jackets. The ropes are mechanically anchored through steel collars, avoiding the use of impregnating resins.

2. Resin impregnated fiber sheet and textile reinforced mortar confining techniques

In columns confined by fiber reinforced sheets impregnated by polymers (fiber reinforced polymers, FRP) an abrupt fracture of the confining reinforcement is usually observed after a certain level of imposed lateral deformations. Figure 1 presents the experimental results of a large experimental program involving high elastic modulus carbon FRP sheet confinement of concrete [4], [9]. In figure 1 the ultimate axial strains range among 0.4% and 2.4% for the specific tests. After the FRP fracture an explosive failure and a rather precipitous drop of the bearing load takes place. The higher the elastic modulus of the FRP jacket, the lower its strain at failure and the lower the axial strain at failure of confined concrete for identical confining rigidity. More recent experimental efforts concern large deformability FRP confining materials made of PET that may present a fracture at far higher levels of concrete axial strains up to 8.5% [10].

Textile Reinforced Mortar (TRMs) jackets are a reliable alternative to FRP jackets with comparable strength and ductility enhancement while they present a higher resistance to elevated temperature [1]. The TRM confinement presents a more gradual failure due to progressive fracture of individual fiber bundles that lead to softening stress-strain behaviour.

Both techniques involve the use of reinforcements working inside a matrix. The epoxy resin or mortar matrix have to be cured for a certain period of time and under controlled environmental conditions in order to bear the final redesign loads. The following sections present a novel strengthening technique utilizing the advantages of the fiber rope reinforcements.

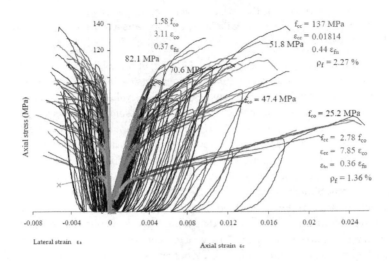

Figure 1. Axial stress versus axial and lateral strain curves of 45 concrete cylinders confined by high-E-modulus carbon FRP sheets under monotonic or cyclic loading. Five concrete strengths and three different volumetric rations are presented.

3. Resin or mortar free PPFR & VFR confinement

In the first part of the experimental program, the vinylon and polypropylene fiber ropes confine standard concrete cylinders (150 mm diameter and 300 mm height) in three different confinement volumetric ratios. The vinylon fiber rope (VFR) is a three-stranded Z-twisted one with 12.66 mm² structural area. Its modulus of elasticity is 15.9 GPa and 4.6% tensile failure strain [2]. Two different polypropylene ropes are used in the research (manufactured by Thrace Plastic Co. S.A.). The first rope is braided, having eight strands (bPPFR) and a very low E-modulus of 2250 MPa. The tensile failure strain is 18%. The structural area of the bPPFR is 21.25 mm². The second polypropylene fiber rope is a two-stranded Z-twisted one (tPPFR) with a structural area of 12.09 mm². Its E-modulus is 1991 MPa and shows 20.35% tensile failure strain. The application is resin-free. Thus, a mechanical anchoring of the rope is applied without the use of impregnating resin. The application of one or multiple rope layers follows a careful wrapping by hand, yet exerting an adequate and continuous tension on the rope. The wrapping process is acceptable only if the confinement is tight enough. Loose wrapping reduces significantly the efficiency of FR wrapping. However thorough wrapping by hand and suitable mechanical anchorage may provide an efficient lateral restriction to concrete. The applications of the two different types of PPFR with different knitting and overall structural diameter reveal no difference in the structural behaviour. The braided multi-strand rope leaves higher percentage of gaps among multiple rope layers than the twisted two-strand rope of half structural area. In addition, the tPPFRs require lower hand force during application in order to achieve an adequate continuous tension. Yet, the braided multi-strand rope provides similar structural effectiveness with the twisted two-strand rope. The concrete strength of the specimens during the tests was 15.56 MPa. Figure 2a presents a concrete column confined by vinylon fiber ropes.

The testing includes monotonic and cyclic axial loading of the columns. The research assesses the effectiveness of rope composite reinforcements considering the whole axial stress versus axial and lateral strain behaviour, as well as the failure values resulting from monotonic or cyclic compressive loading of concrete cylinders. The results are shown in figures 3& 4. Confinement of concrete through fiber ropes leads to substantial upgrade of concrete strain at failure reaching values of 13% strain (around 39 strain ductility). The strength enhancement reached a number of 6.6 times that of plain concrete. Those measured values are recorded from specimen confined by VFR that could not be tested up to failure due to loading machine limitations. Similar is the case with the PPFR confined specimens which reach axial strain ductility of 35.5. All PPFR confined specimens and the heavy VFR confined columns present a failure mode that involves the extensive disintegration of concrete and no fracture of the confining reinforcement occurs. The concrete core disintegrates uniformly up to the point that the transverse restriction by the FR surpasses the transverse rigidity of the concrete. After that point, full unloading of the concrete core does not lead to equilibrium with the external confinement and thus to stable remaining (plastic) axial and lateral strains. On the contrary, the extensively and uniformly cracked core squeezes under the external lateral pressure and the plastic axial and lateral strains of the previous cycle reduce significantly. The specimen lengthens in the axial direction and may become higher than its original height. That unique behaviour of concrete reveals a remarkable efficiency of rope reinforcements to redistribute lateral restriction. That type of concrete response is referred as a "spring-like" behaviour (Figure 2b). Moreover, PPFR show a very low sensitivity to local damage. Both ropes can be easily coupled through simple knots or simple steel chucks. Thus, both ropes can be easily reused even if fractured. More details on the first part of the experiments can be found in [11].

(a)   (b)

Figure 2. Measurement of lateral deformation on concrete and on FR of specimen VinL1V2 (a). Specimen DPPL1p1 after failure of the concrete core during full decompression (b). Adapted from [11].

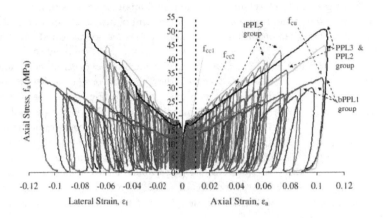

Figure 3. Comparative stress –strain diagrams of PPFR confined concrete. Adapted from [11].

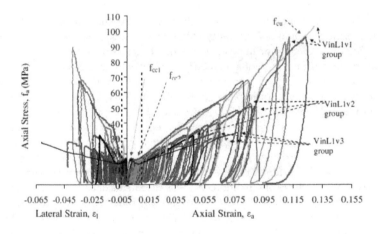

Figure 4. Comparative stress –strain diagrams of VFR confined concrete. Adapted from [11].

Since the majority of specimens could not be tested up to the fracture of the FR, the assessment of the structural behaviour of the columns utilizes the provided lateral rigidity of the confinement $E_l = 0.5\ k_e\ \varrho_{FR}\ E_{FR}$, where E_{FR} is the modulus of elasticity of the FR, ϱ_{FR} is the volumetric ratio and k_e is the effectiveness coefficient due to spiral clear spacing s' that is $k_e = 1-s'/2d$ where d is the section diameter. Thus, the confinement aims at achieving a limited load drop (lower than 20%) from f_{cc1} to f_{cc2} (see Figures 3 and 4) upon full activation of the FR (or no drop at all). Then an ever increasing bearing load response up to failure occurs (stress f_{cu}).

The unique "spring – like" behaviour of concrete at ultimate offers experimental evidence that further support the earlier findings on the requirement of further elaboration and development of experimental data on plastic deformations and more sophisticated modeling approaches sensitive to the path-dependent deformation evolution of confined concrete [5, 12, 13, 14 among else]. Finite element analyses may prove necessary in order to model and generate reliably the geometric nonlinearities and material interactions [12, 15 among others]. The response of confined concrete depends on the mode of loading and on the evolution of axial to lateral stresses during loading and unloading as well. The fiber rope confinement application reveals that the redistribution capacity and minimum sensitivity of the rope reinforcement homogenizes the cracks developed inside the concrete core and the response of the member as whole. Thus, the utilization of the concrete deformability under triaxial compression is rather optimized when vinylon or polypropylene FR materials are used. High deformability materials present low modulus of elasticity and thus the required structural thickness is high. Fiber ropes (FR) can provide a solution and at the same time exclude the use of impregnation resins.

4. Hybrid FRP and FR Confinement

This part of the experimental study concerns the utilization of the unique, ductile behaviour of concrete confined by high deformability elastic materials. Such materials may ensure the reserve of the strain ductility of the concrete under compression, often required due to an overloading in the critical region of a member caused by extreme seismic excitations. In columns confined by FRP sheets an abrupt fracture of the confining reinforcement is usually observed after a certain level of imposed lateral deformations (Figure 1). According to the first part of the experimental research, high deformability elastic confining materials can maintain the integrity of low strength concrete to a remarkable strain level even higher than 10% while presenting ever increasing axial load capacity (rather hardening behaviour). Thus, the study further investigates the confining effects of hybrid glass FRP – polypropylene (PP) fiber rope (FR) external confinement. The tests include concrete cylinders in two series with concrete qualities of C16 and C20 under repeated axial compression cycles of increasing displacement similar to the first part of the experimental study. The specimens are confined externally with only one layer of glass FRP sheet and adequate FR confinement in different volumetric ratios. Herein the presentation is limited to the beneficial effects of the dual action of the glass FRP and PPFR.

The whole experimental program includes 21 specimens of two different concrete strengths, C16 batch with average concrete strength of 25.1 MPa and C20 with concrete strength of 33.7 MPa. Three specimens in each batch are confined by 5 layers of tPPFR to examine the efficiency of FR confinement in columns with varying concrete strength. Another two columns in each batch are confined by 1 layer of glass FRP (GFRP). The glass FRP is of S&P G90/10 type (S&P—Sintecno 1999 [16]) with 300 mm width. Details on the application requirements of the sheet FRP on non-circular columns as well can be found in [7]. The glass sheet has structural thickness of 0.154 mm per layer, tensile modulus of elasticity of 73 GPa and strain at failure equal to 0.028 (after impregnation and curing). Finally, two columns in each batch are confined by both 1 layer of GFRP sheet and 3 layers of PP fiber ropes. Those four columns are constructed in four phases. At first, the external surface of the cylinders is treated with an epoxy paste (P103, Sintecno) in order to fill the undesirable pores and cavities. The full hardening of the paste may require up to 15 hours. After at least 6 hours a layer of the two-component primer S2W resin (Sintecno) is applied on the external surface of the column. The primer resin is left to harden at least 1 hour. After the curing of the primer resin the application of the glass sheet follows with the use of the two-component impregnating resin S2WV (Sintecno). An overlap of 150 mm of the sheet layer prevents anchorage debonding failure. Full development of the impregnating resin capacity may require up to 7 days. After the hardening of the second resin, the 3 layers of the Z-twisted, two-strand polypropylene fiber rope (tPPFR) are applied by hand. The continuous rope may include several ropes reused or not, connected with simple knots if necessary. The FRs are anchored mechanically through steel collars that simply tighten and thus exert an adequate pressure on the ropes against the concrete surface. Thus the anchorage of the ropes is performed mainly by friction. That set up is necessary given the small size of the standard cylinders. In real size concrete components, the anchorage of the rope may be easily constructed outside the region of the expected damage by simply tighten the rope itself with a simple knot or with a suitable steel collar. In concrete specimens confined exclusively by PPFR no special treatment of the external concrete surface is necessary. Consequently, the application of the FR confinement may be directly applied and operating since it may involve zero materials' curing time. Considering the high performance of the rope against local concrete damage and stress redistribution, the FR may be applied on any concrete surface even after severe cracking of the concrete core. However, in confinement applications, large concave surfaces in the external surface should be avoided since they reduce the effectiveness. The following paragraphs present the experimental results of the columns with the dual action of the glass FRP and PPFR confinement.

The typical test setup is presented in Figure 5. Axial and lateral deformations are measured through four displacement meters (linear variable displacement transducers –LVDTs). An advanced laser meter measures the deformation of the rope on the outer surface. The columns are subjected to multiple close compression – decompression – recompression cycles of increasing deformations.

Figures 6 and 7 present the specimen of C16 concrete quality confined by 1 layer of GFRP and 3 layers of tPPFR after the end of the test. It should be mentioned that no tPPFR fracture occurs. After the first fracture of GFRP the axial load drops gradually as it is bared mainly by the tPPFR confinement while the GFRP jacket merely contributes through the interface friction with the adjacent PPFR. At successive cycles of compression the load stabilizes under the PPFR confining action and rises again. During those cycles multiple fractures of the GFRP sheet may develop. Such multiple FRP fractures may be evidenced after the removal of the FR (Figures 6 and 7). Those multiple fractures are characteristic of hybrid confinement and reveal a further utilization of the GFRP sheet as surface reinforcement. All the tests of specimens with hybrid confinement ended prematurely due to loading machine capacity restrictions or after steel collars dislocation and concrete unstable behaviour.

Figure 5. Test setup of specimens confined by 1 layer of GFRP and 3 layers of tPPFR material.

Figure 6. Specimen of C16 concrete confined by 1 layer of GFRP and 3 layers of tPPFR material after the removal of the fiber rope. The GFRP sheet jacket is fractured in two opposite positions. View of the fracture near the overlap region.

Figure 7. Specimen of C16 concrete confined by 1 layer of GFRP and 3 layers of tPPFR material after the removal of the fiber rope. View of the GFRP fracture far from the overlap region.

Since the PPFR confined column present no rope fracture, the remaining response milestones are discussed. The first milestone involves specimens with hybrid confinement that present axial load regaining after the GFRP fracture. The second milestone concerns the further development of axial load equal to the one during first GFRP fracture. The second milestone clearly reveals the efficiency and the effects of the rope confinement.

Figure 8 shows the typical structural response of axial stress versus strain for the GFRP confined specimen of C16 concrete batch, the respective column with 5 layers of tPPFR and the one with hybrid confinement by 1 layer of GFRP and 3 layers of tPPFR. The confinement with 5 layers of PPFR is designed in order to ensure the controlled temporary load drop upon full activation of the PPFR confinement. The lateral rigidity of the PPFR confinement is around 300 MPa. The confinement by 1 layer of GFRP provides a lateral rigidity of around 150 MPa. Thus, the hybrid confinement is designed in order to provide a combined lateral rigidity almost equivalent to that of 5 layers of tPPFR. The GFRP jacket and 3 layers of PPFR correspond to a lateral rigidity of around 340 MPa.

The experimental evidence show that the 3 layers of PPFR lead to further utilization of the GFRP jacket. The fracture of the GFRP occurs at higher axial load and axial strain in hybrid confinement. After the fracture of the GFRP, the outer PPFR confinement results in a temporary and smooth load drop while the GFRP confined specimen presents an abrupt failure. The PPFR confinement can bear the abrupt energy release after the fracture of the FRP and redistributes the resulting uneven lateral pressure through the friction between the FR and sheet confinement. The removed rope presents no fracture or even limited local individual fibers' damage after the end of the test. The load is stabilized under the restrictive action of the PPFR and a load regaining occurs (see the hybrid glass FRP-PPFR confinement in Figure 8). Proper design of hybrid confinement by FRP sheets and high deformability FR as the outermost reinforcement, utilizes fully the confining effects of the FRP sheet up to its fracture. Then adequate FR ensures further increased strain ductility of concrete which withstands also high axial loads and avoids an abrupt load capacity loss. The axial load capacity loss in FRP or in steel stirrup confinement occurring after the fracture of those materials may be replaced by a smooth softening, followed by a hardening behaviour of no "detected failure" for practical applications in structural concrete members.

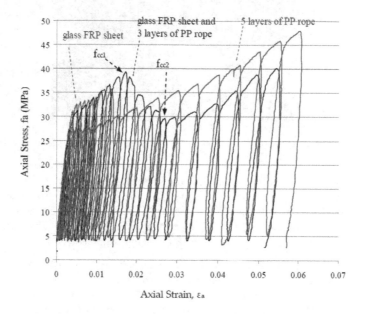

Figure 8. Axial stress versus axial strain curves for concrete specimens confined by 1 layer of glass FRP sheet, by 5 layers of PP rope or by hybrid FRP and 3 layers of PP rope reinforcement.

The temporary load drop can be controlled within desirable levels. As depicted in Figure 8, three layers of PPFR can lead to a minimum axial load before regaining that is higher than the strength of plain concrete.

As mentioned above, the fracture of the GFRP occurs at higher load and strain than the specimen without external PPFR strengthening because of the dual confinement effect and twofold E_l. Figure 8 depicts that the confining effects of GFRP and PPFR confinement are different. The hybrid confinement presents significantly higher load enhancement than the 5 layers of PPFR – with almost equal confinement rigidity – at the same strain levels until the fracture of the GFRP jacket.

Concerning the deformability of concrete, the glass FRP confined column fails at around 1.3% axial strain as the sheet fractures abruptly. On the other hand, an adequate additional quantity of PP fiber rope can provide a hardening stress-strain response of concrete including a temporary load drop with softening behaviour. The deformability goes beyond 5.5%, while no abrupt failure occurs as the PPFR does not fracture. The loading is ended early in all tests due to concrete unstable behaviour or steel collars dislocation.

5. Fiber rope confinement modeling and design

The available confinement models predict the stress-strain behaviour or the ultimate stress and strain values of concrete at the fracture of the confining means. Since the case of hybrid GFRP and PPFR confinement does not include the fracture of the PPFR for the recorded tests, this approach need to be reconsidered. In this section two significant recent existing models are used to predict the stress and strain values at the characteristic point of the fracture of the GFRP sheet due to dual hybrid confinement.

In their study, Teng et al. (2009) [17] propose empirical relations for the strength and strain of FRP confined columns that recognize the confinement stiffness and the FRP tensile to concrete compressive strain ratio as significant parameters. The strength is predicted by the relation:

$$\left.\begin{array}{l} \dfrac{f_{cc}}{f_{co}} = 1 + 3.50 \cdot (\rho_{\kappa} - 0.01) \cdot \rho_{\varepsilon}, \ when\, \rho_{\kappa} \geq 0.01 \\[2mm] \dfrac{f_{cc}}{f_{co}} = 1, \ when\, \rho_{\kappa} < 0.01 \end{array}\right], \ \rho_{\kappa} = \dfrac{2 \cdot E_j \cdot t_j}{\dfrac{f_{co}}{\varepsilon_{co}} \cdot d}, \ \rho_{\varepsilon} = \dfrac{\varepsilon_{je}}{\varepsilon_{co}}, \ \varepsilon_j = 0.586 \cdot \varepsilon_{fu}$$

The strain at failure is given by the relation:

$$\dfrac{\varepsilon_{cc}}{\varepsilon_{co}} = 1,75 + 6,50 \cdot \rho_{\kappa}^{0,80} \cdot \rho_{\varepsilon}^{1,45}, \ \rho_{\kappa} = \dfrac{2 \cdot E_{FRP} \cdot t}{\dfrac{f_{co}}{\varepsilon_{co}} \cdot D}, \ \rho_{\varepsilon} = \dfrac{\varepsilon_{FRP}}{\varepsilon_{co}}$$

In the research by Rousakis et al. (2012) [18], a strength model is proposed that identifies the normalized axial rigidity of the confining means ($\varrho_f E_f / f_{co}$) as a significant parameter. The lateral strain at failure of concrete is found strongly dependent on the modulus of elasticity of its reinforcing fibers (E_f) and on the confinement effectiveness coefficient (k_1). The model has the following form (see also [9], [5] and [6]):

$$f_{cc}/f_{co} = 1 + k_1(f_{le}/f_{co}) = 1 + k_1(0.5\rho_f E_f \varepsilon_{je}/f_{co}) = 1 + (\rho_f E_f / f_{co})(\,0.5\,k_1 \varepsilon_{je}),$$

that is

$$f_{cc}/f_{co} = 1 + (\rho_f E_f / f_{co})(\alpha E_f 10^{-6}/E_{f\mu} + \beta)$$

with $\varrho_f = 4t_f / d$ and $E_{f\mu} = 10$ MPa (for units' compliance). For FRP sheet wraps $\alpha = -0.336$ and $\beta = 0.0223$. For FRP tube encased concrete $\alpha = -0.2300$ and $\beta = 0.0195$.

The above models are applied to the experimental results of GFRP confined columns and of columns with dual GFRP-PPFR confinement. Figures 9 and 10 present the predictions of strength and strain at failure of the two models. The absolute error of the predictions for both strength and strain is less than 7.1% for the columns with hybrid confinement.

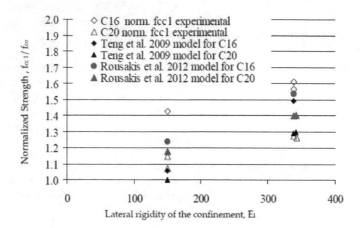

Figure 9. Normalized bearing stress f_{cc1} at first drop of the load to the plain concrete strength versus lateral rigidity of the confinement for 1 layer of glass FRP sheet or by hybrid FRP and 3 layers of PP rope reinforcement for C16 and C20 concrete batches.

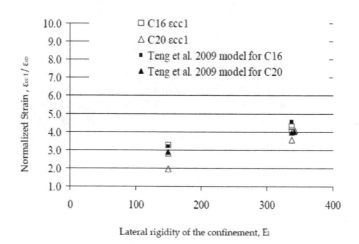

Figure 10. Normalized strain at first drop of the load to the plain concrete strain versus lateral rigidity of the confinement for 1 layer of glass FRP sheet or by hybrid FRP and 3 layers of PP rope reinforcement for C16 and C20 concrete batches.

6. Conclusions

The research presents a novel hybrid confining technique that involves FRP jacketing and fiber ropes mechanically anchored through steel collars. Fiber rope confinement is a "directly applied and operating" strengthening technique. The additional polypropylene fiber rope confinement may enhance the axial stress and strain of concrete prior to FRP fracture. It also restricts the lateral strain of concrete. After the fracture of the FRP, the PPFR restricts the abrupt load drop and stabilizes the concrete softening response up to load regaining and re-hardening. PPFR withstands the abrupt energy release and the multiple fractures of the FRP jacket throughout the loading. No new load drop or PPFR fracture or local PP fiber damage occurs up to axial strains equal to 5.5%. That hybrid technique can enhance remarkably the performance of lightly FRP confined columns that are expected to present abrupt failures during an event of overloading due to seismic excitations.

The redistribution of stress and strain of the rope is feasible because, as mentioned above, the polypropylene fiber rope (PPFR) is not used with resin. The PPFR is applied after the full curing of the common FRP sheets impregnated by resins. Thus the PPFR is bond free when wrapped around the FRP jacket.

In hybrid confining schemes the PPFR is not in conduct with concrete. Thus the concrete is protected by the FRP sheet. The PPFR (if fully wrapped) is expected to provide some protection for the FRP sheet. However no investigations exist on this mater so far. The polypropylene is already used widely as mass fiber reinforcement inside concrete. Thus, even if in direct conduct with concrete, the PPFR exhibits no alkalinity related degradation. On the other hand exposure to UV light or high temperatures should degrade the PPFR. Thus a UV light protecting and fire-resistant finishing mortar is required as in common FRPs applications.

Acknowledgements

Author would like to acknowledge the contribution of Zarras S.A. for providing concrete, of S&P and Sintecno S.A. for providing the glass sheet and the resins, and of Thrace Plastics Co. S.A. for providing the polypropylene fiber ropes. Also, thanks are owed to undergraduate students Sourbati A., Karavelas D., Pekas E., Dimitriadou T., Gouma M. and Anezakis M. for their help in the experimental programs and to DUTh RC lab staff.

Author details

Theodoros C. Rousakis[*]

Address all correspondence to: trousak@civil.duth.gr

Lecturer, Democritus University of Thrace (D.U.Th.), Civil Engineering Department, Engineering Structures Section, Reinforced Concrete Lab, Greece

References

[1] Bournas, B. A., Lontou, P. V., Papanicolaou, C. G., & Triantafillou, T. C. (2007). Textile-Reinforced Mortar (TRM) versus FRP Confinement in Reinforced Concrete Columns. *ACI Structural Journal*, 104(6), 740-748.

[2] Shimomura, T., & Phong, N.H. (2007). Structural Performance of Concrete Members Reinforced with Continuous Fiber Rope. In: Triantafillou T.C. (ed.) FRPRCS-8 Conference: Fiber-Reinforced Polymer Reinforcement for Concrete Structures, July 16-18 University of Patras, Patras, Greece , 408.

[3] Peled, A. (2007). Confinement of Damaged and Nondamaged Structural Concrete with FRP and TRC Sleeves. *Journal of Composites for Construction, ASCE*, 11(5), 514-522.

[4] Rousakis, T.C. (2001). Experimental investigation of concrete cylinders confined by carbon FRP sheets, under monotonic and cyclic axial compressive load. *Research Report, Chalmers University of Technology* [44], Publ.01:2, Work, Göteborg, Sweden.

[5] Rousakis, T. C., Karabinis, A. I., Kiousis, P. D., & Tepfers, R. (2008). Analytical modelling of Plastic Behaviour of Uniformly FRP Confined Concrete Members. *Elsevier, Journal of Composites Part B: Engineering*, 39(7-8), 1104-1113.

[6] Rousakis, T. C., & Karabinis, A. I. (2008). Substandard Reinforced Concrete Members Subjected to Compression- FRP Confining Effects. *RILEM Materials and Structures, Springer Netherlands*, 41(9), 1595-1611.

[7] Rousakis, T. C., & Karabinis, A. I. (2012). Adequately FRP confined reinforced concrete columns under axial compressive monotonic or cyclic loading. *RILEM Materials and Structures, Springer Netherlands*, 45(7), 957-975.

[8] Matthys, S., Toutanji, H., & Taerwe, L. (2006). Stress-Strain Behavior of Large-Scale Circular Columns Confined with FRP Composites. *Journal of Structural Engineering*, 132(1), 123-133.

[9] Rousakis, T.C. (2005). Mechanical behaviour of concrete confined by composite materials. *PhD Thesis. Democritus University of Thrace, Civil Engineering Department, Xanthi; (in Greek)*.

[10] Dai, JG, Bai, YL, & Teng, J.G. (2011). Behavior and Modeling of Concrete Confined with FRP Composites of Large Deformability. *ASCE Journal of Composites for Construction*, 15(6), 963-973.

[11] Rousakis, T.C. (2012). Confinement of Concrete Columns by Fiber Rope Reinforcements. In: Monti J. (ed.) The 6th International Conference on FRP Composites in Civil Engineering- CICE. Rome 13- 15 of June 2012 (accepted for oral presentation).

[12] Karabinis, A. I., Rousakis, T. C., & Manolitsi, G. (2008). D Finite Element Analysis of Substandard Columns Strengthened by Fiber Reinforced Polymer Sheets. *ASCE Journal of Composites for Construction*, 12(5), 531-540.

[13] Yu, T., Teng, J. G., Wong, Y. L., & Dong, S. L. (2010). Finite element modeling of confined concrete-I: Drucker-Prager type plasticity model. *Engineering Structures*, 32-665.

[14] Jiang, J. F., & Wu, Y. F. (2012). Identification of material parameters for Drucker-Prager plasticity model for FRP confined circular concrete columns. *International Journal of Solids and Structures*, 49(3-4), 445-456.

[15] Papanikolaou, V. K., & Kappos, A. J. (2007). Confinement-sensitive plasticity constitutive model for concrete in triaxial compression. *International Journal of Solids and Structures*, 44-7021.

[16] Scherer, J. (1999). S&P--Sintecno, FRP-polymer fibers in strengthening. *User guide, Brunnen.*

[17] Teng, J. G., Jiang, T., Lam, L., & Luo, Y. Z. (2009). Refinement of a Design-Oriented Stress-Strain Model for FRP-Confined Concrete. *ASCE Journal of Composites for Construction*, 13(4), 269-278.

[18] Rousakis, T.C., Rakitzis, T.D., & Karabinis, A.I. (2012). Design- Oriented Strength Model for FRP Confined Concrete Members. ASCE Composites for Construction, accepted for publication available at http://dx.doi.org/10.1061/(ASCE)CC. 1943-5614.0000295

Theoretical - Practical Aspects in FRP

Chapter 6

Analysis of Nonlinear Composite Members Including Bond-Slip

Manal K. Zaki

Additional information is available at the end of the chapter

1. Introduction

Extensive research has been carried out in recent years on the use of FRP composites in strengthening of RC structures. Concrete elements strengthened with FRP undergo significant improvement of strength, ductility and resistance to electrochemical corrosion. Moreover, strengthening concrete member with FRP has the advantages of decreased installation costs and repairs, less stiffness and weight in comparison with steel. The increase in stiffness of the structural elements is undesirable in seismic prone areas. Structural members can be strengthened with FRP jackets provided along the whole length of the member or in regions of maximum straining actions. FRP strengthening can, also, be provided on one face of the structural member as in the case of stiffening the tension fibers of a beam.

For FRP retrofitting problem, the confinement model describing the behavior of rectangular concrete columns retrofitted with externally bonded FRP material and subjected to axial stress was presented by Chaallal et al. [1]. Other researchers investigated the effect of FRP in seismic strengthening of concrete columns, Tastani and Pantazopoulou [2] and Ozcan et al. [3]. They found that FRP retrofitting remarkably increased the strength and ductility of the strengthened members. Some researchers proposed simplified equations for FRP retrofit design of difficient rectangular columns, Ozcan et al. [4].

Other researchers studied reinforced concrete members externally bonded with FRP fabric using commercial software ANSYS, Kachlakev et al. [5], Li et al. [6]. Yan et al. [7] developed an analytical stress-strain model. Purushotham et al. [8] studied piles in berthing structures under uniaxial bending. Kaba and Mahin [9] presented the concept of fiber method in their refined modeling of RC columns for seismic analysis under uniaxial bending.

Some searches were conducted to the problem of biaxial bending. Bresler [10] and Bernardo [11] studied biaxial bending for unretrofitted short rectangular columns.

At early stage of the use of layered beams, full interaction (perfect bond) was assumed in the design. It was until the mid-fifties that Newmark and his co-authors [12] pointed the influence of partial interaction on the overall elastic behavior of steel-concrete composite beams. They derived the governing equations and solved the equilibrium equations expressed in terms of the axial force. Since then, several studies have been conducted to study the problem of bond-slip, Arizumi et al. [13], Daniel and Crisinel [14], Salari et al. [15]. Gara [16] and Ranzi [17] adopted the displacement based finite element formulation to include the vertical slip. Salari et al. [18] also Valipour and Bradford [19] adopted one-dimensional element force-based element to solve the relevant problem. Other researchers [20] and [21] adopted the mixed-procedure, displacement-based together with force-based, to solve the problem. Moreover, nonlinear geometric effects were introduced to the problem by Girhammar and Gopu [22], Girhammar and Pan [23], Čas et al. [24] and Pi et al. [25]. Krawczyk and the co-authors [26,27], Battini et al. [28] developed a corotational formulation for the nonlinear analysis of composite beams with interlayer slip. Nguyen et al. [29], Sousa et al. [30] implemented a finite element model to solve a composite beam column with interlayer slip.

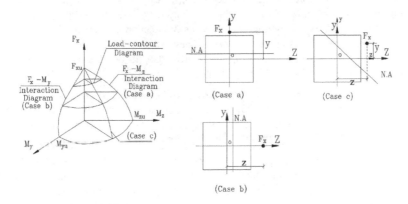

Figure 1. A typical interaction diagram of RC columns

In practice, many RC columns are subjected to biaxial bending. The analysis of such problems are difficult because a trial and adjustment procedure is necessary to find the inclination and depth of the neutral axis. The problem becomes more complicated when the slenderness effect is included. A typical interaction diagram for biaxially loaded column is shown in Fig. 1. Case a and case b are the uniaxial bending about the z axis and y axis respectively. The interaction curves represent the failure envelope for different combinations of the axial load and bending moments. Case c represents the case of a RC column with biaxial bending.

The material nonlinearity is considered to account for concrete cracks and the change of the stress-strain relationship of the different materials. The material nonlinearity is thus introduced by using the FMM together with the incremental iterative solution. The geometric nonlinearity is considered in the present study to account for the deformations occurring due to excess bending moments developed by the effect of axial load. The geometric nonlinearity, thus, considers the slenderness effect of the column. The bond-slip effect is considered by introducing the bond properties of the epoxy resin applied to adhere FRP to the RC column.

The method adopted is accomplished by dividing the column into segments along the member axis to introduce the FEA for the skeletal segments. At each end of the segment, the cross-section is divided into concrete, steel and FRP fibers to introduce the FMM. The properties of a cross-section is calculated by summing up the properties of all the fibers or elemental areas of the particular section. The column segment properties are considered as the average properties of the its end cross-sections. The segment and cross-section discretization are detailed in section 2.

The load is applied incrementally until the maximum allowed strains are reached. An incremental iterative method is employed to solve the problem. After each iteration, the properties of each cross-section are computed according to the material changes occurring and governed by the stress-strain relationship for each material. The properties of each column segment is considered as the average between its end section properties. Those properties are then introduced to the tangential linear stiffness matrix. The geometric nonlinearity is accounted for through the geometric stiffness matrix. Also, the bond-slip effect is considered by the addition of the bond-slip stiffness matrix.

It is, therefore, the aim in this study to adopt the FEA to formulate the linear, geometric and the bond-slip stiffness matrices of composite members subjected to biaxial bending together with axial forces. The model is developed within an updated Lagrangian incremental formulation.

The assumptions of the present analysis are: 1)only longitudinal partial interaction is considered. Axial relative displacement occurs between different elements while the vertical displacement is the same for all elements. 2)small strains and moderate rotations are considered. This assumption represents a rigorous simplification applicable to many problems. 3) Both layers, referred to as elements in the present study, followed the Euler-Bernoulli beam theory. This considers that plane cross-sections remain plane after deformations and perpendicular to the axis of the beam. 4) Shear and torsional deformations are neglected. 5) Effect of the column weight is neglected.

2. Fiber method modelling of frp confined beam columns

The cross-section is divided into concrete, steel and FRP fibers to introduce The FMM is introduced herein to compute the properties of each fiber, thus achieving the properties of the a cross-section by summing up the properties of all its fibers or elemental areas. The meshing is given in Fig. 2(a). The column segment properties are considered as the average properties of its end cross-sections.

The same derivation in the companion paper [31] for columns under biaxial bending is adopted herein after the necessary modifications to solve the column under the effect of slip.

The strain distribution is defined by the maximum compressive strain ε_m, together with the depth of the neutral axis, Z_n. The strains are shown in Fig. 2(b).

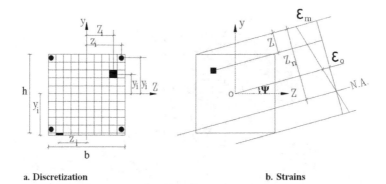

a. Discretization   b. Strains

Figure 2. Cross-section

The following section parameters are then computed

$$\varepsilon_m = \varepsilon_o + \frac{b}{2}\phi_y + \frac{h}{2}\phi_z \tag{1}$$

$$\phi = \sqrt{\phi_y{}^2 + \phi_z{}^2} \tag{2}$$

$$z_n = \frac{\varepsilon_m}{\phi} \tag{3}$$

$$\psi = \tan^{-1}\frac{\phi_z}{\phi_y} \tag{4}$$

where φ_y and φ_z are the curvatures along the y-axis and z-axis respectively and ε_o is the strain at point "O".

The elemental strain is computed as:

$$\varepsilon = \varepsilon_m (1 - \frac{Z}{Z_n}) \tag{5}$$

where Z is the distance from the maximum strain to the element measured perpendicular to the N.A. After determining the strain of each fiber from eq.5, the corresponding elemental modulus of elasticity, G, is determined as detailed in section 4. The elemental properties are computed and summed up to obtain the cross-section properties as shown in the following equations:

$$E A_\alpha = \sum_{i=1}^{nfib} \left[(G_i)\Delta A_i \right] \tag{6a}$$

$$E I_y = \sum_{\alpha=1}^{n} \left[\sum_{i=1}^{nfib} (z_i)^2 (G_i)\Delta A_i \right] \tag{6b}$$

$$E I_z = \sum_{\alpha=1}^{n} \left[\sum_{i=1}^{nfib} (y_i)^2 (G_i)\Delta A_i \right] \tag{6c}$$

$$E I_{yz} = \sum_{\alpha=1}^{n} \left[\sum_{i=1}^{nfib} (y_i)(z_i)(G_i)\Delta A_i \right] \tag{6d}$$

$$(E S_y)_\alpha = \sum_{i=1}^{nfib} \left[(z_i)(G_i)\Delta A_i \right] \tag{6e}$$

$$(E S_z)_\alpha = \sum_{i=1}^{nfib} \left[(y_i)(G_i)\Delta A_i \right] \tag{6f}$$

where α is the counter of an arbitrary element. In the present study, element 1 is the RC section and element 2 is the FRP. n is the total number of elements and is equal to 2 in the present study, i is the counter of fibers, $nfib$ is the total number of fibers of element α, ΔA_i is the area of each fiber, y_i, z_i are distances from the center of the considered fiber to the z and y axes respectively. Those symbols are shown in Fig. 2a. It should be noted that the properties EA, ES_y and ES_z are given separately for each element, while the properties EI_z, EI_y and EI_{yz} are the summation of the corresponding properties of both elements. The reason for this is that the axial displacement is different for each element due to the slip effect while both elements undergo the same curvatures about the z-axis and the y-axis.

3. Displacement-based fiber model with bond-slip

A one dimensional finite element analysis is adopted to solve the column segments. The segments are considered to be of unsymmetric cross-section caused by the inclination of the

N.A. The finite element formulation given by Yang and McGraw [32] to solve the thin-walled, i.e. bare-steel columns, is introduced herein after applying the necessary modification to include the concrete, FRP and bond-slip.

3.1. Displacements and strain fields

The axial displacement of an arbitrary point of an element (α) in the cross-section is given in terms of the displacements of a constant point "c" on the same element as follows

$$u_x = u_{xc\alpha} - zu'_{zc} - yu'_{yc} \tag{7}$$

where $u_{xc\alpha}$ is the axial displacement of the element α, y and z are the vertical and horizontal distances, respectively from the centroid of any follower element α to the centroid of the parent element having $\alpha=1$. In the present study, the concrete section and the FRP are considered to be the parent and the follower elements respectively. And $u'_{yc} and u'_{zc}$ are the derivatives of the transverse displacements u_{yc} and u_{zc}.

y=y_α-y_1 and z=z_α-z_1 however, for simplicity, the reference axes are chosen such that y_1=0 and z_1=0, Fig.3 (a).

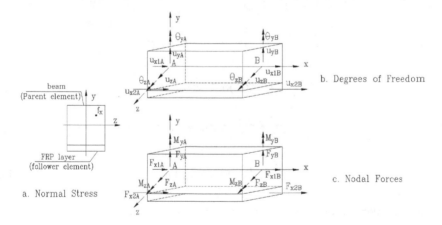

Figure 3. Column Segment

It should be noted that the transverse displacements, u_{yc} and u_{zc}, are considered to be the same for all elements of the beam with respect to the axis of the beam. For simplicity, the symbol "c" is dropped out from the r.h.s. of the equations. The relevant linear strain field can, then, be obtained from the displacement field given by eq. 7 by differentiating the mentioned equation w.r.t. the beam coordinates as

$$\varepsilon_{x\alpha} = (u_{x,x})_{\alpha} = u'_{x\alpha} - yu''_y - zu''_z \tag{8}$$

while the nonlinear strain field is given by

$$\eta_{xx} = \frac{1}{2}(u^2_{x,x} + u^2_{y,x} + u^2_{z,x}) \tag{9}$$

and $u_{x,x}$ is ignored as usual practice. The slip satisfying the compatibility relation with the displacements of element 1 and element 2 is given as

$$u_{xb} = u_{x2} - u_{x1} = u_{x2} - u_{x1} + zu'_z + yu'_y \tag{10}$$

3.2. Forces

In general, the the normal stress f_x shown in Fig. 3 (a) is expressed as

$$f_x = E_\alpha (u'_{x\alpha} - zu''_z - yu''_y) \tag{11}$$

Following the integrations at the cross-section, the stress resultant are expressed as

$$F_x = \int_A f_x dA \quad M_y = \int_A f_x z dA \quad M_z = -\int_A f_x y dA \tag{12}$$

$$F_x = EAu'_x - E(S_y)(u''_z) - E(S_z)(u''_y) \tag{13a}$$

$$M_y = -EI_y u''_z + E(S_y)(u'_x) - E(I_{yz})(u''_y) \tag{13b}$$

$$M_z = EI_z u''_y - E(S_z)(u'_x) + E(I_{yz})(u''_z) \tag{13c}$$

where the elemental properties are previously given in section 2.

3.3. Degrees of freedom and nodal forces

The local coordinates are shown in Fig. 3(b and c).

The displacement vector <u>is given by:

$$<u> = < u_{x1A} \quad u_{yA} \quad u_{zA} \quad \theta_{yA} \quad \theta_{zA} \quad u_{x2A} \quad u_{x1B} \quad u_{yB} \quad u_{zB} \quad \theta_{yB} \quad \theta_{zB} \quad u_{x2B} > \tag{14}$$

where θy and θz are the angles of rotation of the section about y and z axis respectively.

While the vector of the nodal forces $<f>$ is given by:

$$<f>=< F_{x1A}\ F_{yA}\ F_{zA}\ M_{yA}\ M_{zA}\ F_{x2A}\ F_{x1B}\ F_{yB}\ F_{zB}\ M_{yB}\ M_{zB}\ F_{x2B} > \tag{15}$$

3.4. Virtual work

The partial interaction problem is solved using the principle of virtual work. This is accomplished by equating the work of internal stresses to the work of external stresses in an incremental form. The principle of virtual work will be adopted to deduce:

-the linear and nonlinear stiffness matrices of a beam element which consist of an assemblage of two different

elements connected by deformable interface.

-the bond-slip stiffness matrix.

The equilibrium condition for the entire beam-column is then expressed by assembling the vectors and matrices defined for each segment according to the principle of finite elements.

The details are given below.

3.4.1. For the beam with FRP

$$\int_v (E_1 e_{xx} \delta_1 e_{xx})^1 dV + \int_v (f_x \delta_1 \eta_{xx})^1 dV = {}_1^2 R - {}_1^1 R \tag{16}$$

in which ${}_1 e_{xx}$ and ${}_1 \eta_{xx}$ are the linear and non-linear incremental strain respectively, f_x denotes the stress at C_1, E is the modulus of elasticity of the cross-section and ${}^2{}_1 R$ and ${}^1{}_1 R$ are the external virtual work at C_2 and C_1 respectively but both being measured at C_1 and ${}^2{}_2 R$ = the external virtual work expression

substituting equations 8 into eq.16, we get

$$\frac{1}{2}\int_v (E\delta(u^2{}_{,x})dV + \frac{1}{2}\int (f_x\delta(u^2{}_{y,x} + u^2{}_{z,x})dV = {}^2R - {}^1 R \tag{17}$$

A linear displacement field is adopted for the axial displacement, u_x, and a cubic displacement field for other displacements. The incremental displacements are expressed as:

$$u_x = < n_1 > \{\bar{u}_x\}\, u_y = < n_3 > \{\bar{u}_y\}\, u_z = < n_3 > \{\bar{u}_z\} \tag{18}$$

where

$$\langle n_1 \rangle = \langle 1-i \qquad\qquad i \rangle \tag{19a}$$

$$\langle n_3 \rangle = \langle 1-3i^2+2i^3 \qquad i-2i^2+i^3 \qquad 3i^2-2i^3 \qquad i^3-i2 \rangle \tag{19b}$$

in which i is given by the value $\frac{x}{l}$.

The nodal degrees of freedom ie., the ends A and B of the column segment are given by:

$$\{\bar{u}_x\}_a = \langle u_{xA\alpha} \qquad u_{xB\alpha} \rangle \tag{20a}$$

$$\{\bar{u}_y\} = \langle u_{yA} \qquad l\theta_{zA} \qquad u_{yB} \qquad l\theta_{zB} \rangle \tag{20b}$$

$$\{\bar{u}_z\} = \langle u_{zA} \qquad -l\theta_{yA} \qquad u_{zB} \qquad -l\theta_{yB} \rangle \tag{20c}$$

where l is the length of the segment.

3.4.1.1. The linear part

The linear part taken from eq. 17 is

$$\langle \delta u \rangle [K_e]\{u\} = \sum_{\alpha=1}^{n} \left[\frac{1}{2} \int_v (E\delta(u^2_{x,x}) dV \right] \tag{21}$$

in which α is the counter of the considered elements and n is their total number.

$$\langle \delta u \rangle [K_e]\{u\} = \sum_{\alpha=1}^{n} \left[\frac{1}{2} \int_v (E_a \delta(u'_{x\alpha} - yu''_y - zu''_z)^2 dV \right] \tag{22}$$

applying the properties of the cross-section given in eqs 6, the previous expression becomes

$$\langle \delta u \rangle [K_e]\{u\} = \sum_{\alpha=1}^{n} [\frac{1}{2} \int_0^l E_a A_\alpha \delta \left(u'_{x\alpha}\right)^2 dx + \int_0^l (-E_a S_{z\alpha}) \delta(u''_y u'_{x\alpha})^2 dx$$

$$+ \int_0^l (-E_a S_{y\alpha}) \delta(u''_z u'_{x\alpha})^2 dx \tag{23}$$

$$+ \frac{1}{2} \int_0^l E_a I_{z\alpha} \delta \left(u''_y\right)^2 dx + \frac{1}{2} \int_0^l (E_a I_{y\alpha}) \delta(u''_z)^2 dx + \int_0^l (E_a I_{yz\alpha}) \delta(u''_y u''_z) dx]$$

Substituting the interpolation functions in eq. 18, the following equation applies

$$\prec \delta u \succ [K_e]\{u\} = \prec \delta u_{x\alpha} \succ \left[\sum_{\alpha=1}^{n} \int_0^1 \frac{E_\alpha A_\alpha}{l} \{n'_1\} \prec n'_1 \succ di \right]\{u_{x\alpha}\} + \prec \delta u_z \succ \left[\sum_{\alpha=1}^{n} \int_0^1 \frac{E_\alpha I_{y\alpha}}{l^3} \{n''_3\} \prec n''_3 \succ di \right]\{u_z\}$$

$$+ \prec \delta u_y \succ \left[\sum_{\alpha=1}^{n} \int_0^l (\frac{-E_\alpha S_{z\alpha}}{l^2})\{n''_3\} \prec n'_1 \succ di \right]\{u_{x\alpha}\} + \prec \delta u_{x\alpha} \succ \left[\sum_{\alpha=1}^{n} \int_0^l (\frac{-E_\alpha S_{z\alpha}}{l^2})\{n'_1\} \prec n''_3 \succ di \right]\{u_y\}$$

$$+ \prec \delta u_z \succ \left[\sum_{\alpha=1}^{n} \int_0^l (\frac{-E_\alpha S_{y\alpha}}{l^2})\{n''_3\} \prec n'_1 \succ di \right]\{u_{x\alpha}\} + \prec \delta u_{x\alpha} \succ \left[\sum_{\alpha=1}^{n} \int_0^l (\frac{-E_\alpha S_{y\alpha}}{l^2})\{n'_1\} \prec n''_3 \succ di \right]\{u_z\} \tag{24}$$

$$+ \prec \delta u_z \succ \left[\sum_{\alpha=1}^{n} \int_0^l (\frac{E_\alpha I_{y\alpha}}{l^3})\{n''_3\} \prec n''_3 \succ di \right]\{u_z\} + \prec \delta u_y \succ \left[\sum_{\alpha=1}^{n} \int_0^l (\frac{E_\alpha I_{z\alpha}}{l^3})\{n''_3\} \prec n''_3 \succ di \right]\{u_y\}$$

$$+ \prec \delta u_z \succ \left[\sum_{\alpha=1}^{n} \int_0^l (\frac{E_\alpha I_{yz\alpha}}{l^3})\{n''_3\} \prec n''_3 \succ di \right]\{u_y\} + \prec \delta u_y \succ \left[\sum_{\alpha=1}^{n} \int_0^l (\frac{E_\alpha I_{yz\alpha}}{l^3})\{n''_3\} \prec n''_3 \succ di \right]\{u_z\}$$

3.4.1.2. The nonlinear part

The nonlinear part taken from eq. 17 is

$$\prec \delta u \succ [K_g]\{u\} = \frac{1}{2} \int_V f_x [\delta(u'_y)^2 + \delta(u'_z)^2] dV \tag{25}$$

when several elements participate in the nonlinear virtual work, the previous eq becomes

$$\prec \delta u \succ [K_g]\{u\} = \sum_{\alpha=1}^{n} \left[\int_0^l \frac{F_{x\alpha}}{2} [\delta(u'_y)^2 + \delta(u'_z)^2] dx \right] \tag{26}$$

in which α is the counter of the considered elements and is their total number.

$$\prec \delta u \succ [K_g]\{u\} = \prec \delta u_y \succ \left[\int_0^l \sum_{\alpha=1}^{n} \frac{F_{x\alpha}}{l} \{n'_3\} \prec n'_3 \succ di \right]\{u_y\} + \prec \delta u_z \succ \left[\int_0^l \sum_{\alpha=1}^{n} \frac{F_{x\alpha}}{l} \{n'_3\} \prec n'_3 \succ di \right]\{u_z\} \tag{27}$$

when   n=2,   as   in   the   general   case,   then $\left[\int_0^l \sum_{\alpha=1}^{n} \frac{F_{x\alpha}}{l} \{n'_3\} \prec n'_3 \succ di \right]$ becomes

$[\int_0^l \frac{(F_{x1} + F_{x2})}{l} \{n'_3\} \prec n'_3 \succ di]$. The linear and nonlinear stiffness matrices are obtained after performing the integrations in eqs 24 and 27 and are given in the appendix.

3.4.2. For bond-slip

The bond-slip expression given in eq.10 is substituted in the linear portion of the virtual work expression given in eq. 17 and the expression thus becomes

$$\prec \delta u \succ [K_b]\{u\} = \frac{1}{2} \int_v [E_b \delta(-u_{x1} + yu'_y + zu'_z + u_{x2})^2] dV = \prec \delta u \succ [\{^2 f\} - \{^1 f\}] \qquad (28)$$

$$
\begin{aligned}
\prec \delta u \succ [K_b]\{u\} = &\frac{1}{2} \int_0^l E_b A_b \left[\delta(u_{x1})^2 + \delta(u_{x2})^2 - 2\delta(u_{x1} u_{x2}) \right] dx \\
&+ \frac{1}{2} \int_0^l E_b A_b y \left[-2\delta(u_{x1} u'_y) + 2\delta(u_{x2} u'_y) \right] dx \\
&+ \frac{1}{2} \int_0^l E_b A_b z \left[-2\delta(u_{x1} u'_z) + 2\delta(u_{x2} u'_z) \right] dx + \frac{1}{2} \int_0^l E_b A_b y^2 \left[\delta(u'_y)^2 \right] dx \\
&+ \frac{1}{2} \int_0^l E_b A_b z^2 \left[\delta(u'_z)^2 \right] dx + \frac{1}{2} \int_0^l E_b A_b yz \left[2\delta(u'_y u'_z) \right] dx
\end{aligned}
\qquad (29)
$$

$$
\begin{aligned}
\prec \delta u \succ [K_b]\{u\} = &\prec \delta u_{x1} \succ \left[\int_0^l lE_b A_b \{n_1\} \prec n_1 \succ di \right] \{u_{x1}\} + \prec \delta u_{x2} \succ \left[\int_0^l lE_b A_b \{n_1\} \prec n_1 \succ di \right] \{u_{x2}\} \\
&- \prec \delta u_{x2} \succ \left[\int_0^l lE_b A_b \{n_1\} \prec n_1 \succ di \right] \{u_{x1}\} - \prec \delta u_{x1} \succ \left[\int_0^l lE_b A_b \{n_1\} \prec n_1 \succ di \right] \{u_{x2}\} \\
&- \prec \delta u_{x1} \succ \left[\int_0^l E_b A_b y \{n_1\} \prec n'_3 \succ di \right] \{u_y\} - \prec \delta u_y \succ \left[\int_0^l E_b A_b y \{n'_3\} \prec n_1 \succ di \right] \{u_{x1}\} \\
&+ \prec \delta u_{x2} \succ \left[\int_0^l E_b A_b y \{n_1\} \prec n'_3 \succ di \right] \{u_y\} + \prec \delta u_y \succ \left[\int_0^l E_b A_b y \{n'_3\} \prec n_1 \succ di \right] \{u_{x2}\} \\
&- \prec \delta u_{x1} \succ \left[\int_0^l E_b A_b z \{n_1\} \prec n'_3 \succ di \right] \{u_z\} - \prec \delta u_z \succ \left[\int_0^l E_b A_b z \{n'_3\} \prec n_1 \succ di \right] \{u_{x1}\} \\
&+ \prec \delta u_{x2} \succ \left[\int_0^l E_b A_b z \{n_1\} \prec n'_3 \succ di \right] \{u_z\} + \prec \delta u_z \succ \left[\int_0^l E_b A_b z \{n'_3\} \prec n_1 \succ di \right] \{u_{x2}\} \\
&+ \prec \delta u_y \succ \left[\int_0^l \frac{E_b y^2}{l} \{n'_3\} \prec n'_3 \succ di \right] \{u_y\} + \prec \delta u_z \succ \left[\int_0^l (\frac{E_b z^2}{l}) \{n'_3\} \prec n'_3 \succ di \right] \{u_z\} \\
&+ \prec \delta u_z \succ \left[\int_0^l \frac{E_b zy}{l} \{n'_3\} \prec n'_3 \succ di \right] \{u_y\} + \prec \delta u_y \succ \left[\int_0^l (\frac{E_b zy}{l}) \{n'_3\} \prec n'_3 \succ di \right] \{u_z\}
\end{aligned}
\qquad (30)
$$

where A_b is the area of the FRP per unit length of the segment.

The bond-slip matrix is obtained after performing the integrations in eq 30 and is given in the appendix.

Eqs.(24, 27 and 30) can be combined to give:

$$\prec \delta u_{x1} \succ \left[\int_0^l \frac{E_1 A_1}{l}\{n_1'\} \prec n_1' \succ di \right]\{u_{x1}\} + \prec \delta u_{x1} \succ \left[\int_0^l (\frac{-E_1 S_{z1}}{l^2})\{n_1'\} \prec n_3'' \succ di \right]\{u_y\}$$

$$+ \prec \delta u_{x1} \succ \left[\int_0^l (\frac{-E_1 S_{y1}}{l^2})\{n_1'\} \prec n_3'' \succ di \right]\{u_z\}$$

$$+ \prec \delta u_{x1} \succ \left[\int_0^l l E_b A_b \{n_1\} \prec n_1 \succ di \right]\{u_{x1}\} - \prec \delta u_{x1} \succ \left[\int_0^l l E_b A_b \{n_1\} \prec n_1 \succ di \right]\{u_{x2}\} \qquad (31a)$$

$$- \prec \delta u_{x1} \succ \left[\int_0^l E_b A_b z \{n_1\} \prec n_3' \succ di \right]\{u_z\} - \prec \delta u_{x1} \succ \left[\int_0^l E_b A_b y \{n_1\} \prec n_3' \succ di \right]\{u_y\}$$

$$= \prec \delta u_{x1} \succ [\prec {}^2 F_{x1A} \quad {}^2 F_{x1B} \succ^T - \prec {}^1 F_{x1A} \quad {}^1 F_{x1B} \succ^T]$$

$$\prec \delta u_y \succ \left[\int_0^l (\frac{-E_1 S_{z1}}{l^2})\{n_3''\} \prec n_1' \succ di \right]\{u_{x1}\} + \prec \delta u_y \succ \left[\int_0^l (\frac{-E_2 S_{z2}}{l^2})\{n_3''\} \prec n_1' \succ di \right]\{u_{x2}\}$$

$$+ \prec \delta u_y \succ \left[\sum_{\alpha=1}^n \int_0^l (\frac{E_\alpha I_{z\alpha}}{l^3})\{n_3''\} \prec n_3'' \succ di \right]\{u_y\} + \prec \delta u_y \succ \left[\sum_{\alpha=1}^n \int_0^l (\frac{E_\alpha I_{yz\alpha}}{l^3})\{n_3''\} \prec n_3'' \succ di \right]\{u_z\}$$

$$+ \prec \delta u_y \succ \left[\int_0^l \sum_{\alpha=1}^n \frac{F_{x\alpha}}{l}\{n_3'\} \prec n_3' \succ di \right]\{u_y\}$$

$$- \prec \delta u_y \succ \left[\int_0^l E_b A_b y \{n_3'\} \prec n_1 \succ di \right]\{u_{x1}\} + \prec \delta u_y \succ \left[\int_0^l E_b A_b y \{n_3'\} \prec n_1 \succ di \right]\{u_{x2}\} \qquad (31b)$$

$$+ \prec \delta u_y \succ \left[\int_0^l \frac{E_b y^2}{l}\{n_3'\} \prec n_3' \succ di \right]\{u_y\} + \prec \delta u_y \succ \left[\int_0^l (\frac{E_b zy}{l})\{n_3'\} \prec n_3' \succ di \right]\{u_z\}$$

$$= \prec \delta u_y \succ [\prec {}^2 F_{yA} \quad \frac{{}^2 M_{zA}}{l} \quad {}^2 F_{yB} \quad \frac{{}^2 M_{zB}}{l} \succ^T - \prec {}^1 F_{yA} \quad \frac{{}^1 M_{zA}}{l} \quad {}^1 F_{yB} \quad \frac{{}^1 M_{zB}}{l} \succ^T]$$

$$\prec \delta u_z \succ \left[\int_0^l (\frac{-E_1 S_{y1}}{l^2})\{n_3''\} \prec n_1' \succ di \right]\{u_{x1}\} + \prec \delta u_z \succ \left[\int_0^l (\frac{-E_2 S_{y2}}{l^2})\{n_3''\} \prec n_1' \succ di \right]\{u_{x2}\}$$

$$+ \prec \delta u_z \succ \left[\sum_{\alpha=1}^n \int_0^l (\frac{E_\alpha I_{y\alpha}}{l^3})\{n_3''\} \prec n_3'' \succ di \right]\{u_z\} + \prec \delta u_z \succ \left[\sum_{\alpha=1}^n \int_0^l (\frac{E_\alpha I_{yz\alpha}}{l^3})\{n_3''\} \prec n_3'' \succ di \right]\{u_y\}$$

$$+ \prec \delta u_z \succ \left[\int_0^l \sum_{\alpha=1}^n \frac{F_{x\alpha}}{l}\{n_3'\} \prec n_3' \succ di \right]\{u_z\}$$

$$- \prec \delta u_z \succ \left[\int_0^l E_b A_b z \{n_3'\} \prec n_1 \succ di \right]\{u_{x1}\} + \prec \delta u_z \succ \left[\int_0^l E_b A_b z \{n_3'\} \prec n_1 \succ di \right]\{u_{x2}\} \qquad (31c$$

$$+ \prec \delta u_z \succ \left[\int_0^l \frac{E_b z^2}{l}\{n_3'\} \prec n_3' \succ di \right]\{u_z\} + \prec \delta u_z \succ \left[\int_0^l (\frac{E_b zy}{l})\{n_3'\} \prec n_3' \succ di \right]\{u_y\}$$

$$= \prec \delta u_z \succ [\prec {}^2 F_{zA} \quad \frac{-{}^2 M_{yA}}{l} \quad {}^2 F_{zB} \quad \frac{-{}^2 M_{yB}}{l} \succ^T - \prec {}^1 F_{zA} \quad \frac{-{}^1 M_{yA}}{l} \quad {}^1 F_{zB} \quad \frac{-{}^1 M_{yB}}{l} \succ^T]$$

$$\prec \delta u_{x2} \succ \left[\int_0^l \frac{E_2 A_2}{l} \{n_1'\} \prec n_1' \succ di \right] \{u_{x2}\} + \prec \delta u_{x2} \succ \left[\int_0^l (\frac{-E_2 S_{z2}}{l^2}) \{n_1'\} \prec n_3'' \succ di \right] \{u_y\}$$

$$+ \prec \delta u_{x2} \succ \left[\int_0^l (\frac{-E_2 S_{y2}}{l^2}) \{n_1'\} \prec n_3'' \succ di \right] \{u_z\}$$

$$+ \prec \delta u_{x2} \succ \left[\int_0^l l E_b A_b \{n_1\} \prec n_1 \succ di \right] \{u_{x2}\} - \prec \delta u_{x2} \succ \left[\int_0^l E_b A_b y \{n_1\} \prec n_3' \succ di \right] \{u_y\} \qquad (31d)$$

$$- \prec \delta u_{x2} \succ \left[\int_0^l E_b A_b z \{n_1\} \prec n_3' \succ di \right] \{u_z\} - \prec \delta u_{x2} \succ \left[\int_0^l l E_b A_b \{n_1\} \prec n_1 \succ di \right] \{u_{x1}\}$$

$$= \prec \delta u_{x2} \succ [\prec {}^2F_{x2A} \qquad {}^2F_{x2B} \succ^T - \prec {}^1F_{x2A} \qquad {}^1F_{x2B} \succ^T]$$

And upon simplification, the equilibrium equations (31a to 31d) are written in the form

$$[K_e]\{u\} + [K_g]\{u\} + [K_b]\{u\} = \{{}^2f\} - \{{}^1f\} \qquad (32)$$

in which $[K_e]$, $[K_g]$ and $[K_b]$ are the linear, geometric and bond-slip stiffness matrices respectively, $\{u\}$ is the incremental displacement vector and $\{{}^1f\}$ and $\{{}^2f\}$ are the segment nodal forces at the beginning and the end of the incremental step.

The very simple form of the equilibrium equations is

$$[K_t]\{u\} = \{f\} \qquad (33)$$

in which

$$[K_t] = [K_e] + [K_g] + [K_b] \qquad (34)$$

The given procedure can be applied to problems with complete bond by combining elemental properties of the elements 1 and 2 and dropping out the bond-slip stiffness matrix. In this case each $[K]$ will be of order 10*10 instead of 12*12.

4. Stress-strain curves

The constitutive relations for concrete, steel, FRP and bond are schemetically shown in Fig.4.

4.1. Stress strain relationship for FRP

The stress-strain relationship for FRP is considered linear as shown in Fig 4 a

the incremental stress-strain relationship is given by

$$\delta f_f = G_f \delta \varepsilon_f \qquad (35)$$

where G_f is the elemental FRP modulus of elasticity and is expressed as

$$G_f = \frac{f_{fu}}{\varepsilon_{fu}} = E_f \quad \text{when} \quad 0 \le \varepsilon_f \le \varepsilon_{fu} \qquad (36)$$

in which f_f and ε_f are the FRP stress and strain respectively, f_{fu} and $\varepsilon_{f\,u}$ are the ultimate FRP stress and strain respectively and E_f is the modulus of elasticity of FRP.

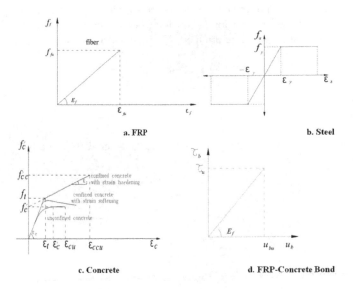

Figure 4. Stress-strain curves

4.2. Stress strain relationship for steel

For simplicity, the stress-strain relationship for the steel bars is considered to be an elastic-perfectly plastic curve neglecting steel hardening. The relationship is shown in Fig. 4.b.

The incremental stress-strain relationship is

$$\delta f_s = G_s \delta \varepsilon_s \qquad (37)$$

where G_s is the incremental steel modulus of elasticity and is expressed as

$$G_s = \frac{f_y}{\varepsilon_y} = E_s \qquad\qquad\qquad \text{when} - \varepsilon_y \leq \varepsilon_s \leq \varepsilon_y \qquad\qquad (38a)$$

$$G_s = 0 \qquad\qquad\qquad \text{when} \varepsilon_s > \varepsilon_y \, or \, \varepsilon_s < -\varepsilon_y \qquad\qquad (38b)$$

in which f_s and ε_s are the steel stress and strain respectively, f_y and ε_y are the yield stress and yield strain respectively and E_s is the modulus of elasticity of steel.

4.3. Stress strain relationship for concrete

For unconfined concrete, the relationship adopted by Al-Noury and Chen [33] was chosen to express the first portion of the compressive stress-strain curve for concrete as a third-degree polynomial. The second portion is considered to be perfectly plastic as shown in Fig. 4.c. The incremental stress-strain relationships is expressed as:

$$\delta f_c = G_c \delta \varepsilon_c \qquad\qquad (39)$$

where

$$G_c = \frac{f_c'}{\varepsilon_c'}\gamma_1 + 2\frac{f_c'}{\varepsilon_c'^2}(3 - 2\gamma_1)\varepsilon_c + 3\frac{f_c'}{\varepsilon_c'^3}(\gamma_1 - 2)\varepsilon_c^2 \quad \text{when} \ \ 0.0 < \varepsilon_c \leq \varepsilon_c' \qquad (40a)$$

$$G_c = 0.0 \qquad \text{when} \ \ \varepsilon_c > \varepsilon_c' \qquad\qquad (40b)$$

in which

$$\gamma_1 = \frac{E_c \varepsilon_c'}{f_c'} \ , \quad \varepsilon_c' = 0.002 \quad \text{and} \quad E_c = 30000\sqrt{f_c'} \qquad\qquad (41)$$

E_c= modulus of elasticity of concrete computed in t/m^2 while f_c' and ε_c' are the maximum unconfined concrete compressive strength and the corresponding strain respectively.

The stress-strain behavior of FRP-confined concrete is largely dependent on the level of FRP confinement. The bilinear stress-strain relationship suggested by Wu et al. [34] is shown in Fig. 4(c) and is adopted herein. The stress-strain curve of concrete confined with sufficient FRP displays a distinct bilinear curve with a second ascending branch as shown in Fig. 4(c). A minimum ratio of FRP confinement strength to unconfined concrete compressive strength f_l/f_c of approximately 0.08 is provided to ensure an ascending second branch in the stress-strain curve. Confinement modulus (E_1) and confinement strength (f_l) are considered to be the two main factors affecting the performance of FRP-confined columns. The two factors are given as:

$$E_1 = \frac{1}{2}\rho_f E_f \tag{42a}$$

$$f_1 = \frac{1}{2}\rho_f f_f \tag{42b}$$

where ρ_f is the volumetric ratio of FRP to concrete, which can be determined for a rectangular section as to a circular section with an equivalent diameter taken as the length of the diagonal of the rectangular section as follows:

$$\rho_f = \frac{4nt_f}{\sqrt{h^2 + b^2}} \tag{43}$$

where h and b are the bigger and smaller dimensions of the cross-section respectively, n is the number of FRP layers and t_f is the thickness of each layer.

The maximum FRP-confined concrete compressive strength and the ultimate axial strain of the FRP-confined concrete compressive stress-strain are given by Rocca et al. [35] as

$$f_{cc}' = f_c' + 3.3k_a f_1 \tag{44a}$$

$$\varepsilon_{ccu} = \varepsilon_c'(1.5 + 12k_b \frac{f_1}{f_c'}(\frac{\varepsilon_{fe}}{\varepsilon_c'})^{0.45}) \le 0.01 \tag{44b}$$

where k_a and k_b are efficiency factors that account for the geometry of the cross-section. In the case of rectangular columns, they depend on the effectively confined area ratio A_e/A_c and the side-aspect ratio h/b. These factors are given by the following expressions:

$$k_a = \frac{A_e}{A_c}\left(\frac{b}{h}\right)^2 \tag{45a}$$

$$k_b = \frac{A_e}{A_c}\left(\frac{b}{h}\right)^{0.5} \tag{45b}$$

$$\frac{A_e}{A_c} = \frac{1 - ((b/h)(h-2r)^2 + (h/b)(b-2r)^2)/(3A_g) - \rho_g}{1 - \rho_g} \tag{46}$$

where A_g is the total cross-sectional area, ρ_g is the ratio of the longitudinal steel reinforcement to the cross-sectional area of a compression member and r is the corner radius of the cross-section.

The slope of the second branch E_2 is computed from the following equation considering the intercept of the second portion with the stress axis equal to f_c' for simplicity, Rocca et al. [35].

$$E_2 = \frac{f_{cc}' - f_c'}{\varepsilon_{ccu}} \tag{47}$$

The transition stress f_t and transition strain ε_t are given by the following equations

$$f_t = (1 + 0.0002E_1)f_c' \tag{48a}$$

$$\varepsilon_t = (1 + 0.0004E_1)\varepsilon_c' \tag{48b}$$

The maximum exerted confining pressure f_{lu} is attained when the circumferential strain in the FRP reaches its ultimate strain ε_{fu} corresponding to a tensile strength f_{fu} [36] and Eq. (42b) becomes

$$f_{lu} = \frac{1}{2}\rho_f f_{fu} = \frac{2f_{fu}nt_f}{\sqrt{h^2+b^2}} = \frac{2nt_f E_f \varepsilon_{fe}}{\sqrt{h^2+b^2}} \tag{49}$$

where $\sqrt{h^2+b^2}$ is the equivalent diameter for non-circular cross-sections. The following equations express the elemental modulus of elasticity for confined concrete in terms of strain.

The effective strain ε_{fe} is computed as the product of an efficiency factor K_e and the ultimate FRP tensile strain ε_{fu}. The factor K_e is to account for the difference between the actual rupture strain observed in FRP-confined concrete specimens and the FRP material rupture strain determined from tensile coupon testing, Wu et al. [34]. The factor ranges from 0.55 to 0.61 and is taken 0.586 in this study.

$$G_c = E_c \quad \text{when} \quad 0 \leq \varepsilon_c \leq \varepsilon_t \tag{50a}$$

$$G_c = E_2 \quad \text{when} \quad \varepsilon_t < \varepsilon_c \leq \varepsilon_{ccu} \tag{50b}$$

4.4. Stress strain relationship for FRP-Concrete Bond

The relationship is shown in Fig. 4.d.

The incremental stress-strain relationship is

$$\delta\tau_b = G_b \delta u_b \tag{51}$$

where G_b is the incremental steel modulus of elasticity and is expressed as

$$G_b = \frac{\tau_b}{u_b} = E_b \quad \text{when} \quad 0 \le u_b \le u_{bu} \tag{52}$$

in which τ_b and u_b are the steel stress and strain respectively, τ_b and u_b are the yield stress and yield strain respectively and E_b is the bond elastic stiffness.

5. Steps of solution followed by the developed program

The mixed procedure is utilized to solve the nonlinear problem. This procedure utilizes a combination of the incremental and iterative (Newton-Raphson) schemes. The load is applied incrementally and after each increment successive iterations are performed. Steps of the solution are then introduced.

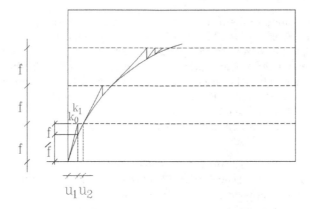

Figure 5. Incremental-iterative method

The mixed procedure is utilized herein to solve the nonlinear problem. This procedure utilizes a combination of the incremental and iterative (Newton-Raphson) schemes. The load is applied incrementally and after each increment successive iterations are performed. The method is illustrated in fig.5.

The combined method is summarized in the following steps:

1. Apply the first increment of load {f} and compute [K_o] assuming no cracks and full bond between the concrete element and the FRP element at the beginning. Compute the displacements {u_1} by solving the equation $[K_o]\{u_1\} = \{f\}$

2.   Compute $[K_1]$ based on the displacement $\{u_1\}$ then compute the load $\{f\}$ from the equation $\left\{f'\right\} = [K_1]\{u_1\}$

3.   Compute $\{\Delta f\}$ as the difference between the applied load $\{f\}$ and the deduced load $\left\{f'\right\}$. Then compute the corresponding displacements $\{u_2\}$ by solving the equation $[K_1]\{u_2\} = \{\Delta f\}$

4.   Repeat steps 2 and 3 until $\{\Delta f\}$ becomes very small.

5.   Repeat all steps again for the next increment.

6. Numerical examples

Two examples are given below. The first example considers a rectangular column fully confined with FRP. Complete bond is considered. The second example is a beam strengthened with FRP on the tension side. In this example the slip between the two elements is considered.

Example 1: The verification of the method is plotted in Fig.6 against experimental results given by Chaallal and Shahawy [1]. The column has across-section of $0.35*0.2$ m^2 and length 2.1m. The concrete has a compression strength 25 MPa and the column is reinforced with 4 grade 60 steel bars of diameter 19 mm each. The steel bars are of 406 MPa yield stress and 206 GPa modulus of elasticity. The specimens are confined with 1mm of carbon fiber reinforced polymer of tensile strength 530 MPa and tensile modulus of elasticity 44 GPa. This gives a confinement ratio, $f_l/f_c' = 0.103$. The present procedure of analysis was adopted to the same specimens and interaction diagrams were plotted. The present results show great accordance with the previous work.

A slight difference in results is observed. It is owed to the provision of corbels in the specimens of Chaallal and Shahawy. They provided large corbels at the ends of the specimens to receive a single load source applied eccentrically thus simulating the combined stress effects in columns. The corbels increased the overall stiffness of the beam column and thus the capacity of loads.

It should be noted that all wraps were characterized by a bidirectional oriented fibers $(0^0/90^0)$ applied along the entire height of the columns. As recommended by ACI 440.2R-02 [37]. The enhancement is only of the significance in members where compression failure is the controlling mode Nanni [38]. This strength enhancement is due to the confining effect of the FRP. When the column is subjected to axial load F_x and moment M_z such that their coordinates lie below the balanced point, the column is considered to be unconfined. This is owed to the limited value of F_x which is considered insufficient to dilate the concrete in the hoop direction thus failing to activate the FRP wrapping effect to confine the concrete. In the present analysis where the wraps are of bidirectional fibers, the point of pure bending is computed accounting for the FRP in the longitudinal direction and its contribution to the flexural capacity according to ACI440.2R-02 [37]. This case was also set by Chaallal and Sha-hawy [1]. Fig.7shows the plots of the column subjected to uniaxial bending Mz and My.

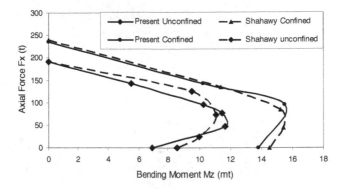

Figure 6. Verification against Shahawy

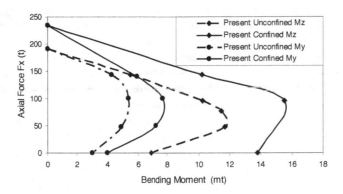

Figure 7. Uniaxial Moments About z and y axes

As expected, the capacity of the about the about the y-axis is less than that about the z-axis. The same model was also subjected to biaxial bending at two axial load levels, namely: Fx=0 and Fx=0.7. The plots of the contour lines of the confined and unconfined columns are given in Figs.(8 and 9).

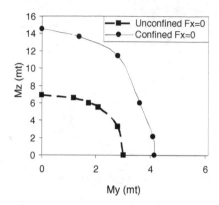

Figure 8. Contour lines (Fx=0.0)

Example 2:The problem of bond-slip was verified against Aprile et al.[39]. A simply supported rectangular beam of length 2.9m loaded by two forces, each=F at the middle
third of the beam. The cross-section is 0.3*0.2 m². Top reinforcement is 226 mm², bottom reinforcement is 339 mm². the beam is strengthened at the bottom by carbon FRP
of width 50 mm and 1.2 mm thickness. The concrete has a compression strength 25
MPa. The steel bars are of 460 MPa yield strength and 210 GPa modulus of elasticity. The carbon fiber reinforced polymer is of tensile strength 2400 MPa and tensile modulus of elasticity 150 GPa. The epoxy resin is of 100 MPa compressive strength and
12.8 GPa modulus of elasticity. The concrete element is considered supported on a roller at one end and hinged at the other end. While the FRP element is considered to
be supported on rollers at both ends. Fig.10 shows the verification of the present analysis if the beam considering bond-slip against Aprile. The curves are plots of the mid-
span deflection of the beam against the applied force (2F). A slight difference is observed
between the two curves. Also, a plot of the reference beam, with no FRP was plotted as reference beam. Another plot of the beam with full bond between the concrete
and the FRP was plotted. At the maximum deflection of the beam with bond-slip, the
reference beam shows nearly 15% decrease in the load capacity while the beam with complete bond achieves nearly 20% increase in the load capacity. In addition, the later
beam undergoes greater deflection and the highest capacity. The curves show two points
of remarkable change in slope indicating remarkable loss of strength in the beam. The
lower point indicates concrete cracking in the middle third of the beam, at the location of the applied concentrated load. The upper point indicates the start of yield of
the bottom steel reinforcement.

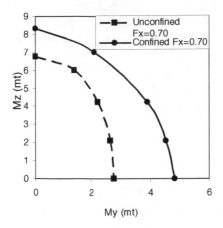

Figure 9. Contour lines (Fx=0.70)

7. Summary and conclusions

The FEA together with the FMM were utilized to solve the problem of RC strengthened with FRP. The structural member solved can be of any slenderness ratio, under any loading and can have any end conditions. The FRP wraps can be totally or partially bonded to the concrete member. The elastic, geometric and bond-slip stiffness matrices of the member in the three-dimension were deduced and given in an appendix.

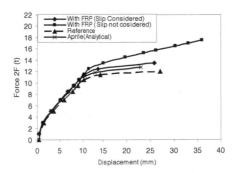

Figure 10. Force-Deflection Diagrams

	u_{x1}	u_y	u_z	θ_y	θ_z	u_{x2}	u_{y1}	u_y	u_z	θ_y	θ_z	u_{x2}
F_{x1}	$\frac{E_1 A_1}{l}$			$\frac{E_1 S_{y1}}{l}$	$\frac{-E_1 S_{z1}}{l}$	$\frac{-E_1 A_1}{l}$				$\frac{-E_1 S_{y1}}{l}$	$\frac{E_1 S_{z1}}{l}$	
F_y		$\frac{12\sum_{a=1}^{n}(E_a I_{za})}{l^3}$	$\frac{12\sum_{a=1}^{n}(E_a I_{yza})}{l^3}$	$\frac{-6\sum_{a=1}^{n}(E_a I_{yza})}{l^2}$	$\frac{6\sum_{a=1}^{n}(E_a I_{za})}{l^2}$			$\frac{-12\sum_{a=1}^{n}(E_a I_{za})}{l^3}$	$\frac{-12\sum_{a=1}^{n}(E_a I_{yza})}{l^3}$	$\frac{-6\sum_{a=1}^{n}(E_a I_{yza})}{l^2}$	$\frac{6\sum_{a=1}^{n}(E_a I_{za})}{l^2}$	
F_z			$\frac{12\sum_{a=1}^{n}(E_a I_{ya})}{l^3}$	$\frac{-6\sum_{a=1}^{n}(E_a I_{ya})}{l^2}$	$\frac{6\sum_{a=1}^{n}(E_a I_{yza})}{l^2}$				$\frac{-12\sum_{a=1}^{n}(E_a I_{ya})}{l^3}$	$\frac{-6\sum_{a=1}^{n}(E_a I_{ya})}{l^2}$	$\frac{6\sum_{a=1}^{n}(E_a I_{yza})}{l^2}$	
M_y				$\frac{4\sum_{a=1}^{n}(E_a I_{ya})}{l}$	$\frac{-4\sum_{a=1}^{n}(E_a I_{yza})}{l}$	$\frac{E_2 S_{y2}}{l}$	$\frac{-E_1 S_{y1}}{l}$	$\frac{6\sum_{a=1}^{n}(E_a I_{ya})}{l^2}$	$\frac{6\sum_{a=1}^{n}(E_a I_{yza})}{l^2}$	$\frac{2\sum_{a=1}^{n}(E_a I_{ya})}{l}$	$\frac{-2\sum_{a=1}^{n}(E_a I_{yza})}{l}$	$\frac{-E_2 S_{y2}}{l}$
M_z					$\frac{4\sum_{a=1}^{n}(E_a I_{za})}{l}$	$\frac{-E_2 S_{z2}}{l}$	$\frac{E_1 S_{z1}}{l}$	$\frac{-6\sum_{a=1}^{n}(E_a I_{za})}{l^2}$	$\frac{-6\sum_{a=1}^{n}(E_a I_{yza})}{l^2}$	$\frac{-2\sum_{a=1}^{n}(E_a I_{yza})}{l}$	$\frac{2\sum_{a=1}^{n}(E_a I_{za})}{l}$	$\frac{E_2 S_{z2}}{l}$
F_{x2}						$\frac{E_1 A_2}{l}$				$\frac{-E_2 S_{y2}}{l}$	$\frac{E_2 S_{z1}}{l}$	$\frac{-E_1 A_2}{l}$
F_{x1}							$\frac{E_1 A_1}{l}$			$\frac{E_1 S_{y1}}{l}$	$\frac{-E_1 S_{z1}}{l}$	
F_y								$\frac{12\sum_{a=1}^{n}(E_a I_{za})}{l^3}$	$\frac{12\sum_{a=1}^{n}(E_a I_{yza})}{l^3}$	$\frac{6\sum_{a=1}^{n}(E_a I_{yza})}{l^2}$	$\frac{-6\sum_{a=1}^{n}(E_a I_{za})}{l^2}$	
F_z									$\frac{12\sum_{a=1}^{n}(E_a I_{ya})}{l^3}$	$\frac{6\sum_{a=1}^{n}(E_a I_{ya})}{l^2}$	$\frac{-6\sum_{a=1}^{n}(E_a I_{yza})}{l^2}$	
M_y										$\frac{4\sum_{a=1}^{n}(E_a I_{ya})}{l}$	$\frac{-4\sum_{a=1}^{n}(E_a I_{yza})}{l}$	$\frac{E_2 S_{y2}}{l}$
M_z											$\frac{4\sum_{a=1}^{n}(E_a I_{za})}{l}$	$\frac{-E_2 S_{z2}}{l}$
F_{x2}												$\frac{E_2 A_2}{l}$

Appendix 1. Linear Stiffness Matrix [K_e]

	u_{x1}	u_y	u_z	θ_y	θ_z	u_{x2}	u_{y1}	u_y	u_z	θ_y	θ_z	u_{x2}
F_{x1}												
F_y		$\frac{6}{5l}$			$\frac{1}{10}$			$\frac{-6}{5l}$			$\frac{1}{10}$	
F_z			$\frac{6}{5l}$	$\frac{-1}{10}$					$\frac{-6}{5l}$	$\frac{-1}{10}$		
M_y				$\frac{2l}{15}$					$\frac{1}{10}$	$\frac{-l}{30}$		
M_z					$\frac{2l}{15}$			$\frac{-1}{10}$			$\frac{-l}{30}$	
F_{x2}												
F_{x1}												
F_y								$\frac{6}{5l}$			$\frac{-1}{10}$	
F_z									$\frac{6}{5l}$	$\frac{1}{10}$		
M_y										$\frac{2l}{15}$		
M_z											$\frac{2l}{15}$	
F_{x2}												

All the elements of the geometric stiffness matrix are multiplied by the factor $\sum_{a=1}^{n} F_{xa}$ which in the general case when $a=2$, the previous term is ($F_{x1}+F_{x2}$).

Appendix 2. Nonlinear Stiffness Matrix [K_g].

	u_{x1}	u_y	u_z	θ_y	θ_z	u_{x2}	u_{x1}	u_y	u_z	θ_y	θ_z	u_{x2}
F_{x1}	$\frac{lE_bA_b}{3}$	$\frac{E_bA_by}{2}$	$\frac{E_bA_bz}{2}$	$\frac{lE_bA_bz}{12}$	$\frac{-lE_bA_by}{12}$	$\frac{-lE_bA_b}{3}$	$\frac{lE_bA_b}{6}$	$\frac{-E_bA_by}{2}$	$\frac{-E_bA_bz}{2}$	$\frac{-lE_bA_bz}{12}$	$\frac{lE_bA_by}{12}$	$\frac{-lE_bA_b}{6}$
F_y		$\frac{6E_bA_by^2}{5l}$	$\frac{6E_bA_byz}{5l}$	$\frac{-E_bA_byz}{10}$	$\frac{E_bA_by^2}{10}$	$\frac{-E_bA_by}{2}$	$\frac{E_bA_by}{2}$	$\frac{-6E_bA_by^2}{5l}$	$\frac{-6E_bA_byz}{5l}$	$\frac{-E_bA_byz}{10}$	$\frac{E_bA_by^2}{10}$	$\frac{-E_bA_by}{2}$
F_z			$\frac{6E_bA_bz^2}{5l}$	$\frac{-E_bA_bz^2}{10}$	$\frac{E_bA_byz}{10}$	$\frac{-E_bA_bz}{2}$	$\frac{E_bA_bz}{2}$	$\frac{-6E_bA_byz}{5l}$	$\frac{-6E_bA_bz^2}{5l}$	$\frac{-E_bA_bz^2}{10}$	$\frac{E_bA_byz}{10}$	$\frac{-E_bA_bz}{2}$
M_y				$\frac{2lE_bA_bz^2}{15}$	$\frac{-2lE_bA_byz}{15}$	$\frac{-lE_bA_bz}{12}$	$\frac{-lE_bA_bz}{12}$	$\frac{E_bA_byz}{10}$	$\frac{E_bA_bz^2}{10}$	$\frac{-lE_bA_bz^2}{30}$	$\frac{lE_bA_bzy}{30}$	$\frac{lE_bA_bz}{12}$
M_z					$\frac{2lE_bA_by^2}{15}$	$\frac{lE_bA_by}{12}$	$\frac{lE_bA_by}{12}$	$\frac{-E_bA_by^2}{10}$	$\frac{-E_bA_byz}{10}$	$\frac{lE_bA_byz}{30}$	$\frac{-lE_bA_by^2}{30}$	$\frac{-lE_bA_by}{12}$
F_{x2}						$\frac{lE_bA_b}{3}$	$\frac{-lE_bA_b}{6}$	$\frac{E_bA_by}{2}$	$\frac{E_bA_bz}{2}$	$\frac{lE_bA_bz}{12}$	$\frac{-lE_bA_by}{12}$	$\frac{lE_bA_b}{6}$
F_{x1}							$\frac{lE_bA_b}{3}$	$\frac{-E_bA_by}{2}$	$\frac{-E_bA_bz}{2}$	$\frac{lE_bA_bz}{12}$	$\frac{-lE_bA_by}{12}$	$\frac{-lE_bA_b}{3}$
F_y			Symmetric					$\frac{6E_bA_by^2}{5l}$	$\frac{6E_bA_byz}{5l}$	$\frac{E_bA_byz}{10}$	$\frac{-E_bA_by^2}{10}$	$\frac{E_bA_by}{2}$
F_z									$\frac{6E_bA_bz^2}{5l}$	$\frac{E_bA_bz^2}{10}$	$\frac{-E_bA_byz}{10}$	$\frac{E_bA_bz}{2}$
M_y										$\frac{2lE_bA_bz^2}{15}$	$\frac{-2lE_bA_byz}{15}$	$\frac{-lE_bA_bz}{12}$
M_z											$\frac{2lE_bA_by^2}{15}$	$\frac{lE_bA_by}{12}$
F_{x2}												$\frac{lE_bA_b}{3}$

Appendix 3. Bond-slip Stiffness Matrix $[K_b]$

Two examples were studied. The first example considers a rectangular column fully confined with FRP. Complete bond was considered. Contour lines can be plotted at any load level. The second example is a beam strengthened with FRP on the tension side. In this example the slip between the two elements was considered. Load-deflection diagrams show that there exist two points of drop in stiffness, the first is due to concrete cracking under the concentrated loads and the second is due to the yield of steel. Extensive research is required to study the effect of the aspect ratio of the concrete cross-section, the strength of the concrete, the strength of FRP, the thickness of FRP and the properties of the epoxy resin used.

Author details

Manal K. Zaki[*]

Address all correspondence to: manalzaki64@yahoo.com.

Department of Civil and Construction Engineering, Higher Technological Institute, 6th October Branch, Guiza Egypt

References

[1] Challal, O., Shahawy, M., & Hassan, M. (2003). Performance of axially loaded short rectangular columns strengthened with carbon fiber-reinforced polymer wrapping. *J Comp Const, ASCE*, 7(3), 200-208.

[2] Tastani, S. P., & Pantazopoulou, S. J. (2008). Detailing procedures for seismic rehabiliation of reinforced concrete members wiyh fiber reinforced polymers. *Engineering Structures*, 2, 450-461.

[3] Ozcan, O., Binici, B., & Ozceke, G. (2008). Improving seismic performance of dificient reinforced concrete columns using carbon fiber reinfored polymers. *Engineering Structures*, 30(6), 1632-1646.

[4] Ozcan, O., Binici, B., & Ozceke, G. (2010). Seismic strengthening of rectangular reinforced concrete columns using fiber reinfored polymers. *Engineering Structures*, 32(4), 964-973.

[5] Kachlakev, D., Thomas, M., & Yim, S. (2001). Finite element modeling of reinforced concrete structures strengthened with frp laminates. *Report for Oregon Department of Ransportation Salem*.

[6] Li, G. K., Su-Seng, P. S., Helms, J. E., & Stukklefield, M. A. (2003). Investigation into frp repaired RC columns. *J Comp. Struct*, 62, 83-80.

[7] Yan, Z., Pantelides, C. P., & Reaveley, L. D. (2006). Fiber reinforced polymer jacketed and shape-modified compression members: I-experimental behavior. *Struct J, ACI*, 103(6), 885-893.

[8] Purushotham, B. R., Alagusundaramoorthy, P., & Sundaravalivelu, R. (2009). Retrofitting of RC piles using GFRP composites. *Journal of Civil Engineering, KSCE*, 13(1), 39-47.

[9] Kaba, S. A., & Mahin, S. A. (1984). Refined modeling of reinforced concrete columns for seismic analysis. *Report No. UBC/EERC-84/3. Ca: University of California, Berkeley*.

[10] Bresler, B. Design criteria for reinforced columns under axial load and biaxial bending. ACI J. (1960). , 32(5), 481-490.

[11] Bernardo, A. L. (2007). Investigation of biaxial bending of reinforced concrete columns through fiber method modeling. *Journal of Research in Science, Computing and Engineering*, 4(3), 61-73.

[12] Newmark, M. N., Siess, C. P., & Viest, I. M. (1951). Tests and analysis of composite beams with incomplete interaction. *Proceedings of the Society for Experimental Stress Analysis*, 9, 175-92.

[13] Arizumi, Y., Hamada, S., & Kajita, T. (1981). Elastic-plastic analysis of composite beams with incomplete interaction by finite element method. *Comp. Struct*.

[14] Daniel, B. J., & Crisinel, M. (1993). Composite slab behavior and strength analysis. *Part I: Calculation procedure. J. Struct. Engrg.*, ASCE, 119(1), 16-35.

[15] Salari, M. R., Spacone, E., Shing, P. B., & Frangopol, D. M. (1997). Behavior of composite structures under cyclic loading. Build. *To Last, Proc., ASCE Struct.Congr. VX, Kempner Jr. L and Brown CB, eds.*, ASCE, New York, 1.

[16] Gara, F., Ranzi, G., & Leoni, G. (2006). Displacement-based formulations for composite beams with longitudinal slip and vertical uplift. *Internat J Numer Methods Engrg.*, 65(8), 1197-220.

[17] Ranzi, G., Gara, F., & Ansourian, P. (2006). General method of analysis for composite beams with longitudinal and transverse partial interaction. *Computers and Structures.*

[18] Salari, M. R., & Spacone, E. (2001). Finite element formulations of one-dimensional elements with bond-slip. *Eng Struct.*, 23, 815-26.

[19] Valipour, Goudarzi. H., & Bradford, M. M.A.(2012). A new shape function for tapered three-dimensional beams with flexible connections. Journal Of Constructional Steel Research , 70, 43-50.

[20] Dall'Asta, A., & Zona, A. (2004). Three-field mixed formulation for the nonlinear analysis of composite beams with deformable shear connection. *Finite Elem Anal Design*, 40, 425-48.

[21] Ayoub, A., & Filippou, F. C. (2002). Mixed formulation of nonlinear steel-concrete composite beam element. *J Struct Eng*, 126(3), 371-81.

[22] Grihammar, U. A., & Gopu, V. K. A. Composite beam-columns with interlayer slip-Exact analysis. J Struct Eng. ASCE (1993).

[23] Grihammar, U.A.P. (2007). Exact static analysis of partially composite beams and beam-columns. *Int J mech Sci*, 49, 139-55.

[24] Čas, B., Sage, M., & Planinc, I. (2004). Non-linear finite element analysis of composite planar frames with interlayer slip. *Comput Struct.*, 82, 1901-12.

[25] Pi, Y. L., Bradford, M. A., & Uy, B. (2006). second order nonlinear inelastic analysis of composite steel-concrete members. I: Theory. *J Struct Eng.*, ASCE, 132(5), 751-61.

[26] Krawczyk, P., Frey, F., & Zielinsky, A. P. (2007). Large deflections of laminated beams with interlayer slips Part 1: Model development. *Eng comput.*, 24(1), 17-32.

[27] Krawczyk, P., & Rebora, B. (2007). Large deflections of laminated beams with interlayer slips Part 2: finite element development. *Eng comput.*, 24(1), 33-51.

[28] Battini, J. M., Nguyen, Q. H., & Hjiaj, M. (2009). Non-linear finite element analysis of composite beams with interlayer slip. *Comput Struct.*, 87, 904-12.

[29] Nguyen, Q. H., & Hjiaj, M. (2011). Exact finite element model for shear-deformable two-layer beams with discrete shear connection. *Finite Elements in Analysis and Design*, 47, 718-727.

[30] Sousa Jr., J. B. M., Oliveira, C. E. M., & da Silva, A. R. (2010). Displacement-based non-linear finite element analysis of composite beams with partial interaction. *Journal of constructional Steel Research.*, 66, 772-779.

[31] Zaki, M.K. (2011). Investigation of FRP strengthened circular columns under biaxial bending. *Engineering Structures*, 33(5), 1666-1679.

[32] Yang, Y. B., & Mc Graw, W. (1986). Stiffness Matrix for Geometric Nonlinear Analysis. *Journal of Structural Engineering*, ASCE, 112, 853-877.

[33] Al-Noury, S. I., & Chen, W. F. (1982). Behavior and design of reinforced and composite concrete sections. *Journal of Structural Division*, ASCE, 17169, 1266-1284.

[34] Wu, G., Lu, Z. T., & Wu, Z. S. (2006). Strength and ductility of concrete cylinders confined with FRP composites. *Construction and Building Materials*, 20, 134-148.

[35] Rocca, S., Galati, N., & Nanni, A. (2009). Interaction diagram methodology for desgin of FRP-confined reinforced concrete columns. *Construction and Building Materials*, 23, 1508-1520.

[36] Wu, G., Lu, Z. T., & Wu, Z. S. (2003). Stress-strain relationship for FRP-confined concrete cylinders. *Proceedings of the 6th international symposium on FRP reinforcement for concrete structures (FRPRCS), Singapor*, 552-560.

[37] American Concrete Institute. (2002). ACI440.2R, Guide for the design and construction of externally bonded FRP systems for strengthening of concrete structures. *Farmington Hills, MI, USA: American Concrete Institute.*

[38] Nanni, A., & Bradford, N.M. (1995). FRP jacketed concrete under uniaxial compression. *Constr. and Build. Mat.*, 9(2), 115-124.

[39] Aprile, A., Spacone, E., & Limkatanyu, S. (2001). Role of bond in RC beams strengthened with steel and FRP plates. *Journal of Structural Division*, ASCE, 22694, 1445-11452.

Circular and Square Concrete Columns Externally Confined by CFRP Composite: Experimental Investigation and Effective Strength Models

Riad Benzaid and Habib-Abdelhak Mesbah

Additional information is available at the end of the chapter

1. Introduction

The use of fiber reinforced polymers (FRP) jackets as an external mean to strengthen existing RC columns has emerged in recent years with very promising results [1-13], among others. Several studies on the performance of FRP wrapped columns have been conducted, using both experimental and analytical approaches. Such strengthening technique has proved to be very effective in enhancing their ductility and axial load capacity. However, the majority of such studies have focused on the performance of columns of circular cross section. The data available for columns of square or rectangular cross sections have increased over recent years but are still limited. This field remains in its developmental stages and more testing and analysis are needed to explore its capabilities, limitations, and design applicability. This study deals with a series of tests on circular and square plain concrete (PC) and reinforced concrete (RC) columns strengthened with carbon fiber reinforced polymer (CFRP) sheets. According to the obtained test results, FRP-confined specimens' failure occurs before the FRP reached their ultimate strain capacities. So the failure occurs prematurely and the circumferential failure strain was lower than the ultimate strain obtained from standard tensile testing of the FRP composite. In existing models for FRP-confined concrete, it is commonly assumed that the FRP ruptures when the hoop stress in the FRP jacket reaches its tensile strength from either flat coupon tests which is herein referred to as the FRP material tensile strength. This phenomenon considerably affects the accuracy of the existing models for FRP-confined concrete. On the basis of the effective lateral confining pressure of composite jacket and the effective circumferential FRP failure strain a new equations were proposed to predict the strength of FRP-confined concrete and corresponding strain for each of the cross section geometry used, circular and square. The predictions of the proposed equations are shown to agree well with test data. The specimen notations are as follows. The first letter refers to section shape: C for circular and S for square. The next two letters indicate the type of concrete: PC for plain concrete and RC for reinforced concrete, followed by the concrete mixture: I for normal strength (26 MPa), II for medium strength (50 MPa) and III for high strength (62 MPa). The last letters specifies the number of CFRP layers (0L, 1L and 3L), followed by the number of specimen.

2. Observed Behaviour of FRP Confined Concrete

2.1. FRP-Confined Concrete in Circular Columns

The confinement action exerted by the FRP on the concrete core is of the passive type, that is, it arises as a result of the lateral expansion of concrete under axial load. As the axial stress increases, the corresponding lateral strain increases and the confining device develops a tensile hoop stress balanced by a uniform radial pressure which reacts against the concrete lateral expansion [14,15]. When an FRP confined cylinder is subject to axial compression, the concrete expands laterally and this expansion is restrained by the FRP. The confining action of the FRP composite for circular concrete columns is shown in Figure 1.

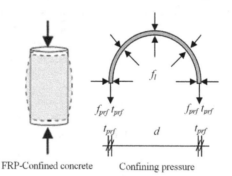

FRP-Confined concrete   Confining pressure

Figure 1. Confinement action of FRP jacket in circular sections

For circular columns, the concrete is subject to uniform confinement, and the maximum confining pressure provided by FRP composite is related to the amount and strength of FRP and the diameter of the confined concrete core. The maximum value of the confinement pressure that the FRP can exert is attained when the circumferential strain in the FRP reaches its ultimate strain and the fibers rupture leading to brittle failure of the cylinder. This confining pressure is given by:

$$f_l = \frac{2t_{frp}E_{frp}\varepsilon_{fu}}{d} = \frac{2t_{frp}f_{frp}}{d} = \frac{\rho_{frp}f_{frp}}{2} \tag{1}$$

Where f_l is the lateral confining pressure, E_{frp} is the elastic modulus of the FRP composite, ε_{fu} is the ultimate FRP tensile strain, f_{frp} is the ultimate tensile strength of the FRP composite, t_{frp} is the total thickness of the FRP, d is the diameter of the concrete cylinder, and ρ_{frp} is the FRP volumetric ratio given by the following equation for fully wrapped circular cross section:

$$\rho_{frp} = \frac{\pi d\, t_{frp}}{\pi d^2/4} = \frac{4 t_{frp}}{d}$$ (2)

2.2. FRP-Confined Concrete in Square Columns

A square column with rounded corners is shown in Figure 2. To improve the effectiveness of FRP confinement, corner rounding is generally recommended. Due to the presence of internal steel reinforcement, the corner radius Rc is generally limited to small values. Existing studies on steel confined concrete [16-18] have led to the simple proposition that the concrete in a square section is confined by the transverse reinforcement through arching actions, and only the concrete contained by the four second-degree parabolas as shown in Figure 2a is fully confined while the confinement to the rest is negligible. These parabolas intersect the edges at 45°. While there are differences between steel and FRP in providing confinement, the observation that only part of the section is well confined is obviously also valid in the case of FRP confinement. Youssef et al. (2007) [19] showed that confining square concrete members with FRP materials tends to produce confining stress concentrated around the corners of such members, as shown in Figure 2b. The reduced effectiveness of an FRP jacket for a square section than for a circular section has been confirmed by experimental results [2,20]. Despite this reduced effectiveness, an FRP-confined square concrete column generally also fails by FRP rupture [9,20]. In Equation (1), d is replaced by the diagonal length of the square section. For a square section with rounded corners, d can be written as:

$$d = \sqrt{2}b - 2Rc\left(\sqrt{2}-1\right)$$ (3)

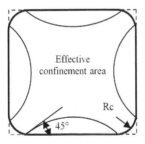

(a) Effectively confined concrete   (b) Dilated square column confined with
    in a square column   carbon/epoxy jacket [19]

Figure 2. Confinement action of FRP composite in square sections

3. Different Behaviour Between Steel and FRP Composite

It is well known that concrete expands laterally before failure. If the lateral expansion is pre-vented, a substantial concrete strength and deformation enhancements may be gained. Thus, the expected enhancement in the axial load capacity of the columns wrapped with FRP may be due to two factors; first: the confinement effect of the externally bonded trans-verse fibers, and second: the direct contribution of longitudinally aligned fibers. Different behaviour between steel and FRP composite was observed due to the stress-strain relation-ship of each material shown in Figure 3. Fiber-reinforced polymer is linear elastic up to final brittle rupture when subject to tension while steel has an elastic-plastic region [21]. This is a very important property in terms of structural use of FRP composite. A part from illustrat-ing typical strength differences between these materials, these curves give a clear contrast between the brittle behaviour of FRP composite and the ductile behaviour of steel. Steel con-finement is based on the same mechanics of FRP. However, a fundamental difference is due to the stress-strain behaviour of steel, which after the initial linearly elastic phase displays the yielding plateau. Therefore, after reaching the maximum value corresponding to the yielding stress, the confinement pressure remains constant (neglecting strain hardening).

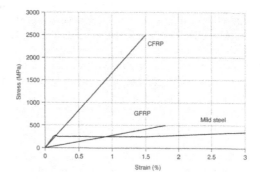

Figure 3. Typical FRP and mild-steel stress-strain curves [21]

4. Experimental Program

4.1. Materials Properties

Concrete mixtures : Three concrete mixtures were used to achieve the desired range of uncon-fined concrete strength (26, 50 and 62 MPa), as shown in Table 1. Mixtures were prepared in the laboratory using a mechanical mixer and were used to cast the concrete specimens which were wrapped with CFRP sheets after drying.

CFRP composites : The carbon-fiber fabric used in this study were the SikaWrap-230C/45 product, a unidirectional wrap. The resin system that was used to bond the carbon fabrics over the specimens in this work was the epoxy resin made of two-parts, resin and hardener. The mixing ratio of the two components by weight was 4:1. SikaWrap-230C/45 was field laminated using Sikadur-330 epoxy to form a carbon fiber reinforced polymer wrap (CFRP) used to strengthen the concrete specimens. The mechanical properties, including the modulus and the tensile strength of the CFRP composite, were obtained through tensile testing of flat coupons. The tensile tests were conducted essentially following the NF EN ISO 527-(1, 2 and 5) recommendations. The tensile specimen configuration is represented in Figure 3a. All of the tests coupons were allowed to cure in the laboratory environment for at least 7 days. Prior to the testing, aluminum plates were glued to the ends of the coupons to avoid premature failure of the coupon ends, which were clamped in the jaws of the testing machine. The tests were carried out under displacement control at a rate of 1mm/min. The longitudinal strains were measured using strain gages at mid-length of the test coupon. The load and strain readings were taken using a data logging system and were stored in a computer. Main mechanical properties obtained from the average values of the tested coupons are summarized below:

- Thickness (per ply) : 1 mm

- Modulus E_{frp} : 34 GPa

- Tensile strength f_{frp} : 450 MPa

- Ultimate strain ε_{fu} : 14 ‰

Note that the tensile strength was defined based on the cross-sectional area of the coupons, while the elastic modulus was calculated from the stress-strain response.

4.2. Fabrication of Test Specimens

The experimental program was carried out on: 1) cylindrical specimens with a diameter of 160 mm and a height of 320 mm; 2) short columns specimens with a square cross section of 140x140 mm and a height of 280 mm. For all RC specimens the diameter of longitudinal and transverse reinforcing steel bars were respectively 12 mm and 8 mm. The longitudinal steel ratio was constant for all specimens and equal to 2.25%.The yield strength of the longitudinal and transversal reinforcement was 500 MPa and 235 MPa; respectively. The specimen notations are as follows. The first letter refers to section shape: C for circular and S for square. The next two letters indicate the type of concrete: PC for plain concrete and RC for reinforced concrete, followed by the concrete mixture: I for normal strength (26 MPa), II for medium strength (50 MPa) and III for high strength (62 MPa). The last letters specifies the number of CFRP layers (0L, 1L and 3L), followed by the number of specimen. Specimens involved in the experimental work are indicated in Table 1.

Specimen designation	Concrete mixture	Nominal dimensions [mm]	Number of CFRP layers	Number of specimens	Unconfined concrete strength [MPa]
CPCI.0L			0	2	
CPCI.1L			1	1	
CPCI.3L	I	Ø160 x 320	3	1	
CRCI.0L			0	2	
CRCI.1L			1	2	
CRCI.3L			3	2	26
SPCI.0L			0	2	
SPCI.1L			1	1	
SPCI.3L	I	140x140x280	3	1	
SRCI.0L			0	2	
SRCI.1L			1	2	
SRCI.3L			3	2	
CPCII.0L			0	2	
CPCII.1L			1	1	
CPCII.3L	II	Ø160 x 320	3	1	
CRCII.0L			0	2	
CRCII.1L			1	2	
CRCII.3L			3	2	50
SPCII.0L			0	2	
SPCII.1L			1	1	
SPCII.3L	II	140x140x280	3	1	
SRCII.0L			0	2	
SRCII.1L			1	2	
SRCII.3L			3	2	
CPCIII.0L			0	2	
CPCIII.1L			1	1	
CPCIII.3L	III	Ø160 x 320	3	1	
CRCIII.0L			0	2	
CRCIII.1L			1	2	
CRCIII.3L			3	2	62
SPCIII.0L			0	2	
SPCIII.1L			1	1	
SPCIII.3L	III	140x140x280	3	1	
SRCIII.0L			0	2	
SRCIII.1L			1	2	
SRCIII.3L			3	2	

Table 1. Details of test specimens

4.3. Fiber-Reinforced Polymer Wrapping

After 28 days of curing, the FRP jackets were applied to the specimens by hand lay-up of
CFRP Wrap with an epoxy resin. The resin system used in this work was made of two parts,
namely, resin and hardener. The components were thoroughly mixed with a mechanical agi-
tator for at least 3 min. The concrete cylinders were cleaned and completely dried before the
resin was applied. The mixed Sikadur-330 epoxy resin was directly applied onto the substrate at a rate of 0,7 kg/m². The fabric was carefully placed into the resin with gloved hands
and smooth out any irregularities or air pockets using a plastic laminating roller. The roller
was continuously used until the resin was reflected on the surface of the fabric, an indication
of fully wetting. After the application of the first wrap of the CFRP, a second layer of resin at
a rate of 0,5 kg/m² was applied on the surface of the first layer to allow the impregnation of
the second layer of the CFRP, The third layer is made in the same way. Finally, a layer of
resin was applied on the surface of wrapped cylinders. This system is a passive type in that
tensile stress in the FRP is gradually developed as the concrete dilates. This expansion is
confined by the FRP jacket, which is loaded in tension in the hoop direction. Each layer was
wrapped around the cylinder with an overlap of ¼ of the perimeter to avoid sliding or debonding of fibers during tests and to ensure the development of full composite strength
(Figure 4). The wrapped cylinder specimens were left at room temperature for 1 week for
the epoxy to harden adequately before testing.

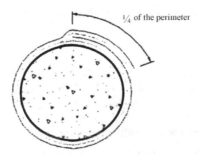

Figure 4. Wrapped cylinder specimens

4.4. Test Procedures

Specimens were loaded under a monotonic uni-axial compression load up to failure. The compressive load was applied at a rate corresponding to 0,24 MPa/s and was recorded with an
automatic data acquisition system. Axial and lateral strains were measured using apprecia-
ble extensometer. The instrumentation included one radial linear variable differential transducers (LVDTs) placed in the form of a hoop at the mid-height of the specimens. Measurement
devices also included three vertical LVDTs to measure the average axial strains. Prior to testing, all CFRP-wrapped cylinders, as well as the plain concrete cylinders, were capped with
sulfur mortar at both ends.The test setup for the cylinders is as shown in Figure 5.

Figure 5. Test set-up

5. Test Results and Discussion

5.1. Overall Behavior

Compression behavior of the CFRP wrapped specimens was mostly similar in each series in terms of stress-strain curves and failure modes of the columns. From the average experimental results reported in Table 2, it can be seen that the increase in strength and axial strain varied according to the unconfined concrete strength, the cross section shape and the amount of confinement provided by CFRP (expressed in number of layers).

The test results described in Table 2 indicate that CFRP-confinemnt can significantly enhance the ultimate strengths and strains of both plain- and RC-columns. As observed for normal-strength RC specimens (26Mpa) with circular and square cross-sections, the average increase in strength were in the order of 69% and 22% over its unconfined concrete strength for columns with 1 layer, 141% and 46% for columns with 3 layers of CFRP jackets, respectively, while the respective values for medium-strength concrete (50 MPa) were 33% and 17% for 1 layer, 72% and 30% for 3 layers of CFRP jackets. Regarding high-strength concrete specimens (62 MPa) with circular and square cross-sections, f'_{co} increased on average 20% and 17% for 1 layer, 50% and 24% for CFRP jackets of 3 layers, respectively.

The axial strains corresponding to CFRP-confined columns (ε_{cc}), for the normal-strength RC specimens with circular and square cross-sections, were on average 4.06 and 1.41 times that of unconfined concrete (ε_{co}) for 1 layer, 6.09 and 1.95 times for 3 layers of CFRP jackets, respectively, while the respective values for medium-strength concrete were 2.76 and 1.32

times for 1 layer, 4.49 and 1.69 times for 3 layers. For high-strength concrete specimens with circular and square cross-sections, ε_{cc} increased 1.39 and 1.03 times for 1 layer, 2.29 and 1.37 times for CFRP jackets of 3 layers, respectively.

Figure 6 shows the increase in compressive strength versus the unconfined concrete strength f_{co} for plain and RC columns confined with one and three layers of CFRP wrap. It is evident that as the unconfined concrete strength increases, the confinement effectiveness decreases. The FRP-wrapped cylinders with the least f_{co} (26 MPa) show the maximum increases in confined strength f'_{cc}. Figure 7 shows the effect of f_{co} on the peak strain ε_{cc} of the confined concrete. Test results clearly showed that the confinement effectiveness reduces with an increase in the unconfined concrete strength for both circular and square columns and strength enhancement was more significant for circular columns than for square ones. This is due to the concentration of stresses at the corner of the square section and consequently to the lower confining pressure and smaller effective confined concrete core area.

Compared to the FRP-confinement-effectiveness, the confinement provided by the minimum transverse reinforcing steel required by Eurocode 2 led to a limited enhancement in both compressive strength and axial strain with respect to plain concrete specimens. With the exception of SRCI.0L specimens, where its presence contributed to a significant increase in the prism load carrying capacity and ductility as shown in Figures 6 and 7.

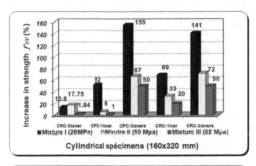

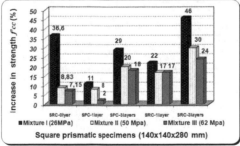

Figure 6. Effect of unconfined strength of concrete on peak stresses

Concrete mixture	Specimen Code	f'_{co} [MPa]	f'_{cc} [MPa]	f'_{cc}/f'_{co}	ε_{cc} [‰]	$\varepsilon_{cc}/\varepsilon_{co}$	$\varepsilon_{h,rup}$ [‰]	$\varepsilon_{h,rup}/\varepsilon_{ho}$
	CPCI.0L		25.93	1.00	2.73	1.00	1.77	1.00
	CPCI.1L	25.93	39.63	1.52	12.78	4.68	13.12	7.41
	CPCI.3L		66.14	2.55	15.16	5.55	13.18	7.44
I (26MPa)	CRCI.0L		29.51	1.00	3.77	1.00	4.95	1.00
	CRCI.1L	29.51	49.88	1.69	15.34	4.06	13.15	2.65
	CRCI.3L		71.35	2.41	22.98	6.09	13.24	2.67
	CPCII.0L		49.46	1.00	1.69	1.00	1.33	1.00
	CPCII.1L	49.46	52.75	1.06	2.52	1.49	2.90	2.18
	CPCII.3L		82.91	1.67	7.27	4.30	13.15	9.88
II (50MPa)	CRCII.0L		58.24	1.00	3.02	1.00	5.05	1.00
	CRCII.1L	58.24	77.51	1.33	8.36	2.76	13.16	2.60
	CRCII.3L		100.41	1.72	13.58	4.49	13.18	2.61
	CPCIII.0L		61.81	1.00	2.64	1.00	2.40	1.00
	CPCIII.1L	61.81	62.68	1.01	3.04	1.15	2.46	1.02
	CPCIII.3L		93.19	1.50	9.80	3.71	12.89	5.37
III (62MPa)	CRCIII.0L		63.01	1.00	2.69	1.00	4.90	1.00
	CRCIII.1L	63.01	76.21	1.20	3.75	1.39	5.20	1.06
	CRCIII.3L		94.81	1.50	6.18	2.29	5.62	1.14
	SPCI.0L		24.77	1.00	2.17	1.00	3.62	1.00
	SPCI.1L	24.77	27.66	1.11	5.58	2.57	12.23	3.37
	SPCI.3L		32.03	1.29	6.05	2.78	13.23	3.65
I (26MPa)	SRCI.0L		33.59	1.00	4.29	1.00	9.38	1.00
	SRCI.1L	33.59	41.02	1.22	6.08	1.41	11.58	1.23
	SRCI.3L		49.12	1.46	8.40	1.95	14.38	1.53
	SPCII.0L		48.53	1.00	3.38	1.00	3.83	1.00
	SPCII.1L	48.53	52.52	1.08	4.03	1.19	7.34	1.91
	SPCII.3L		58.25	1.20	6.72	1.98	9.88	2.57
II (50MPa)	SRCII.0L		52.82	1.00	4.07	1.00	7.50	1.00
	SRCII.1L	52.82	62.04	1.17	5.41	1.32	8.56	1.14
	SRCII.3L		69.09	1.30	6.89	1.69	10.83	1.44
	SPCIII.0L		59.53	1.00	3.56	1.00	3.89	1.00
	SPCIII.1L	59.53	61.30	1.02	3.69	1.03	3.97	1.02
	SPCIII.3L		70.35	1.18	4.94	1.38	6.69	1.71
III (62MPa)	SRCIII.0L		63.79	1.00	3.75	1.00	5.71	1.00
	SRCIII.1L	63.79	74.84	1.17	3.87	1.03	5.74	1.01
	SRCIII.3L		79.59	1.24	5.14	1.37	7.96	1.39

Table 2. Mean-values of experimental results of CFRP-wrapped specimens

5.2. Stress-Strain Response

Representative stress-strain curves for each series of tested CFRP-wrapped specimens are reported in Figure 8 for normal-strength concrete (26 MPa), Figure 9 for medium-strength concrete (50 MPa) and in Figure 10 for high-strength concrete (62 MPa). These figures give the
axial stress versus the axial and lateral strains for circular and square specimens with zero,
one and three layers of CFRP wrap. It can be clearly noticed that both the stress and strain at
failure for the confined specimens were higher than those for unconfned ones. These figures
shows also how the ductility of the concrete specimens was affected by the increase of the
degree of confinement.

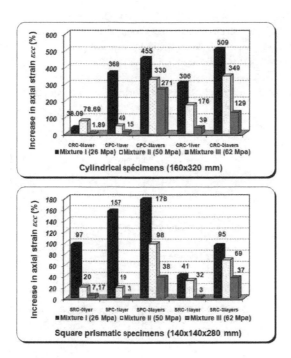

Figure 7. Effect of unconfined strength of concrete on peak strains

The obtained stress-strain curves which characterize the CFRP confined concrete are mostly
bilinear. The first zone is essentially a linear response governed by the stiffness of the unconfined concrete, which indicates that no confinement is activated in the CFRP wraps since the
lateral strains in the concrete are very small. The strengthening effect of the CFRP layers begins only after the concrete has reached the peak strength of the unconfined concrete: transversal strains in the concrete activate the FRP jacket. In this region little increases of load
produce large lateral expansions, and consequently a higher confining pressure. In the case
of circular sections the section is fully confined, therefore the second slope is positive, show-

ing the capacity of confining pressure to limit the effects of the deteriorated concrete core, which allows reaching higher stresses. With this type of stress-strain curves (the increasing type), both the compressive strength and the ultimate strain are reached at the same point and are significantly enhanced. Instead in the cases of square sections (sharp edges) with a small amount of FRP, the peak stress is similar to that of unconfined concrete, indicating the fact that the confining action is mostly limited at the corners, producing a confining pressure not sufficient to overcome the effect of concrete degradation. Otherwise with low levels of confinement (one CFRP layer), the second part of the bilinear curve shifts from strain hardening to a flat plateau, and eventually to a sudden strain softening with a drastically reduced ductility.

From the trends shown in Figures 8, 9 and 10, it is clear that, unlike normal strength concrete, in medium- to high- strength concrete, confining the specimens with one CFRP layer does not significantly change the stress-strain behavior of confined concrete from that of unconfined concrete except for a limited increase in compressive strength. In that case the stress-strain curve terminates at a stress f'_{cu} (stress in concrete at the ultimate strain) $< f'_{co}$, the specimen is said to be insufficiently confined. Such case should not be allowed in design.

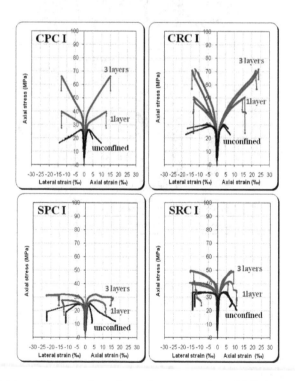

Figure 8. Experimental stress strain curves of normal-strength concrete specimens (26 MPa)

Circular and Square Concrete Columns Externally Confined by CFRP Composite: Experimental
Investigation and Effective Strength Models

203

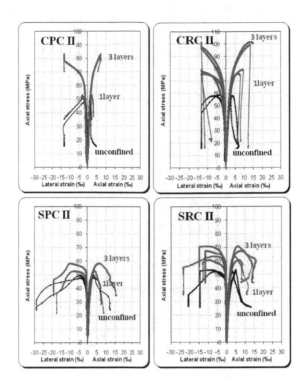

Figure 9. Experimental stress strain curves of medium-strength concrete specimens (50 MPa)

5.3. Failure Modes

Figure 11 illustrate the failure modes for circular and square columns wrapped with CFRP
sheets. All the CFRP-wrapped cylinders failed by the rupture of the FRP jacket due to hoop
tension. The CFRP-confined specimens failed in a sudden and explosive manner and were
only preceded by some snapping sounds. Many hoop sections formed as the CFRP rup-
tured. These hoops were either concentrated in the central zone of the specimen or distributed over the entire height. The wider the hoop, the greater the section of concrete that
remained attached to the inside faces of the delaminated CFRP. Regarding confined concrete
prisms, failure initiated at or near a corner, because of the high stress concentration at these
locations. Collapse occured almost without advance warning by sudden rupture of the composite wrap. For all confined specimens, delamination was not observed at the overlap location of the jacket, which confirmed the adequate stress transfer over the splice.

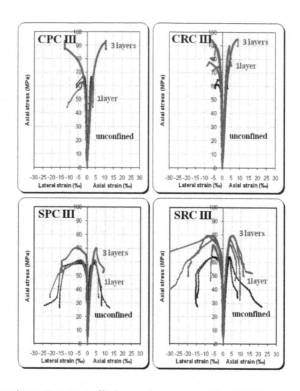

Figure 10. Experimental stress strain curves of high-strength concrete specimens (62 MPa)

Figure 11. Typical failure modes for the tested specimens

6. Model of FRP-Confined Concrete

6.1. Circular Columns

6.1.1. Compressive Strength of FRP-Confined Concrete

Various models for confinement of concrete with FRP have been developed. The majority of these models were performed on plain concrete specimens' tests. A limited number of tests have been reported in the literature on the axial compressive strength and strain of reinforced-concrete specimens confined with FRP. Most of the existing strength models for FRP-confined concrete adopted the concept of Richart et al. (1929) [22], in which the strength at failure for concrete confined by hydrostatic fluid pressure takes the following form:

$$f'_{cc} = f'_{co} + k_1 \cdot f_l \tag{4}$$

Where f'_{cc} and f'_{co} are the compressive strength of confined and the unconfined concrete respectively, f_l is the lateral confining pressure and k_1 is the confinement effectiveness coefficient. In applying their model to steel-confined concrete, Richart et al. (1929) [22] assumed that k_1 is a constant equal to 4.1. However, several studies revealed that existing models for the axial compressive strength of steel-confined concrete are unconservative and cannot be used for FRP-confined concrete (see: [6,21,23-27]; among others). Many authors have raised towards the steel-based confinement models the objection that they do not account for the profound difference in uniaxial tensile stress-strain behavior between steel and FRP. According to these authors, while the assumption of constant confining pressure is still realistic in the case of steel confinement in the yield phase, it cannot be extended to FRP materials which do not exhibit any yielding and therefore apply on the concrete core a continuously increasing inward pressure. However, a number of strength models have been proposed specifically for FRP-confined concrete which employ Equation (4) with modified expressions for k_1 (e.g. [6,7,23-25,27-36]). Most of these models used a constant value for k_1 (between 2 and 3.5) indicating that the experimental data available in the literature show a linear relationship between the strength of confined concrete f'_{cc} and the lateral confining pressure f_l ([7,29,31-37]). Other researchers expressed k_1 in nonlinear form in terms of f_l/f'_{co} or f_l [6,23-25,27,28,30].

FRP Circumferential Failure Strain

According to the obtained test results, cylinder failure occurs before the FRP reached their ultimate strain capacities ε_{fu}. So the failure occurs prematurely and the circumferential failure strain was lower than the ultimate strain obtained from standard tensile testing of the FRP composite. This phenomenon considerably affects the accuracy of the existing models for FRP-confined concrete. Referring to Table 3, for example, the rupture of the low-strength-cylinder IRCC.2.3L corresponded to a maximum composite extension (circumferential failure strain) $\varepsilon_{h,rup}$ of 12.42 ‰ which is lower than the ultimate composite strain ε_{fu} (14

‰) as it represent about 88 % of it. This reduction in the strain of the FRP composites can be attributed to several causes as reported in related literature [6,33,38]:

- The curved shape of the composite wrap or misalignment of fibers may reduce the FRP axial strength;

- Near failure the concrete is internally cracked resulting in non-homogeneous deformations. Due to this non-homogeneous deformations and high loads applied on the cracked concrete, local stress concentrations may occur in the FRP reinforcement.

Concrete mixture	Specimen code	ε_{fu} (‰)	$\varepsilon_{h.rup.}$ (‰)	$\varepsilon_{h.rup.}/\varepsilon_{fu}$
	CRCI.1L.$_1$	14	13.15	0.939
	CRCI.1L.$_2$	14	13.16	0.940
I (26 MPa)	CRCI.3L.$_1$	14	14.06	1.004
	CRCI.3L.$_2$	14	12.42	0.887
	CPCI.1L.$_1$	14	13.12	0.937
	CPCI.3L.$_1$	14	13.18	0.941
	CRCII.1L.$_1$	14	13.17	0.940
	CRCII.1L.$_2$	14	13.16	0.940
II (50 MPa)	CRCII.3L.$_1$	14	13.20	0.942
	CRCII.3L.$_2$	14	13.17	0.940
	CPCII.1L.$_1$	14	2.90	0.207
	CPCII.3L.$_1$	14	13.15	0.939
	CRCIII.1L.$_1$	14	7.79	0.556
	CRCIII.1L.$_2$	14	2.61	0.186
III (62 MPa)	CRCIII.3L.$_1$	14	4.10	0.292
	CRCIII.3L.$_2$	14	7.15	0.510
	CPCIII.1L.$_1$	14	2.46	0.175
	CPCIII.3L.$_1$	14	12.89	0.920

Table 3. Average hoop rupture strain ratios (circular specimens)

Effective FRP Strain Coefficient

In existing models for FRP-confined concrete, it is commonly assumed that the FRP ruptures when the hoop stress in the FRP jacket reaches its tensile strength from either flat coupon tests which is herein referred to as the FRP material tensile strength. This assumption is the basis for calculating the maximum confining pressure f_l (the confining pressure reached when the FRP ruptures) given by Equation (1). The confinement ratio of an FRP-confined specimen is defined as the ratio of the maximum confining pressure to the unconfined concrete strength (f_l/f'_{co}).

However, experimental results show that, the FRP material tensile strength was not reached at the rupture of FRP in FRP-confined concrete. Table 4 provides the average ratios between the measured circumferential strain at FRP rupture ($\varepsilon_{h,rup}$) and the ultimate tensile strain of the FRP material (ε_{fu}). It is seen that, when all circular specimens of the present study are considered together, the average ratio ($\varepsilon_{h,rup}/\varepsilon_{fu}$) has a value closer to 0.73 and is referred to, in this paper, as the effective FRP strain coefficient η. Thus, the maximum confining pressure given by Equation (1) can be considered as a nominal value. The effective maximum lateral confining pressure is given by:

$$f_{l,eff} = \frac{2t_{frp}E_{frp}\varepsilon_{h,rup}}{d} = \frac{2t_{frp}E_{frp}\eta\,\varepsilon_{fu}}{d} = \eta\,f_l \tag{5}$$

Table 3 indicates that the assumption that the FRP ruptures when the stress in the jacket reaches the FRP material tensile strength is invalid for concrete confined by FRP wraps.

Proposed Equation

A simple equation is proposed to predict the peak strength of FRP-confined concrete of different unconfined strengths based on regression of test data reported in Table 4. Figure 12 shows the relation between actual confinement ratio $f_{l,eff}/f'_{co}$ and the strengthening ratio f'_{cc}/f'_{co} for the cylinders of the test series. It can be seen that, strengthening ratio is proportional to the volumetric ratio and the strength of FRP (in terms of effective lateral confining pressure $f_{l,eff}$) and is inversely proportional to unconfined concrete strength. Therefore the relationship may be approximated by a linear function. The trend line of these test data can be closely approximated using the following equation:

$$\frac{f'_{cc}}{f'_{co}} = 1 + 2.20\frac{f_{l,eff}}{f'_{co}} \tag{6}$$

Using a reduction factor η of 0.73 with the replacement of $f_{l,eff}$ by f_l into Equation (6) the ultimate axial compressive strength of FRP-confined concrete takes the form:

$$\frac{f'_{cc}}{f'_{co}} = 1 + 1.60\frac{f_l}{f'_{co}} \tag{7}$$

Figure 13 is a plot of the strengthening ratio f'_{cc}/f'_{co} against the confinement ratio f_l/f'_{co}. The trend line of this figure shows a much greater average confinement effectiveness coefficient k_1. This can be attributed to the effect of the effective lateral confining pressure.

Specimen code	f'_{co} (Mpa)	t_{cfrp} (mm)	E_{cfrp} (Gpa)	ε_{fu} (‰)	$\varepsilon_{h.rup.}$ (‰)	f_l / f'_{co}	$f_{l.eff} / f'_{co}$	f'_{cc} / f'_{co}	ε_{co} (‰)	$\varepsilon_{cc} / \varepsilon_{co}$
CRCI.1L.₁	29.51	1	34	14	13.15	0.201	0.189	1.714	3.77	4.225
CRCI.1L.₂	29.51	1	34	14	13.16	0.201	0.189	1.666	3.77	3.912
CRCI.3L.₁	29.51	3	34	14	14.06	0.604	0.607	2.400	3.77	5.893
CRCI.3L.₂	29.51	3	34	14	12.42	0.604	0.536	2.435	3.77	6.297
CPCI.1L.₁	25.93	1	34	14	13.12	0.229	0.215	1.528	2.73	4.681
CPCI.3L.₁	25.93	3	34	14	13.18	0.688	0.648	2.550	2.73	5.553
CRCII.1L.₁	58.24	1	34	14	13.17	0.102	0.096	1.302	3.02	2.440
CRCII.1L.₂	58.24	1	34	14	13.16	0.102	0.096	1.359	3.02	3.096
CRCII.3L.₁	58.24	3	34	14	13.20	0.306	0.288	1.742	3.02	4.543
CRCII.3L.₂	58.24	3	34	14	13.17	0.306	0.288	1.705	3.02	4.450
CPCII.1L.₁	49.46	1	34	14	2.90	0.120	0.024	1.066	1.69	1.491
CPCII.3L.₁	49.46	3	34	14	13.15	0.360	0.338	1.676	1.69	4.301
CRCIII.1L.₁	63.01	1	34	14	7.79	0.094	0.052	1.237	2.69	1.706
CRCIII.1L.₂	63.01	1	34	14	2.61	0.094	0.017	1.181	2.69	1.081
CRCIII.3L.₁	63.01	3	34	14	4.10	0.283	0.082	1.506	2.69	1.438
CRCIII.3L.₂	63.01	3	34	14	7.15	0.283	0.144	1.503	2.69	3.156
CPCIII.1L.₁	61.81	1	34	14	2.46	0.096	0.016	1.014	2.64	1.151
CPCIII.3L.₁	61.81	3	34	14	12.89	0.288	0.265	1.507	2.64	3.711

Table 4. Data and results of CFRP wrapped cylinders

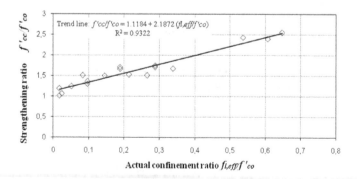

Figure 12. Strengthening ratio vs. actual confinement ratio

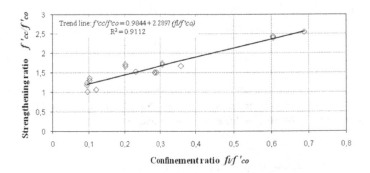

Figure 13. Strengthening ratio vs. confinement ratio

6.1.2. Axial Strain of FRP-Confined Concrete

Early investigation showed that for steel confined concrete, the axial compressive strain ε_{cc} at the peak axial stress can be related to the lateral confining pressure [22] by:

$$\varepsilon_{cc} = \varepsilon_{co}\left(1 + k_2 \frac{f_l}{f'_{co}}\right) \tag{8}$$

Where ε_{co} is the axial strain of the unconfined concrete at its peak stress and k_2 is the strain enhancement coefficient. Richart et al. (1929) [22] suggested $k_2 = 5\ k_1$ for steel-confined concrete. For FRP-confined concrete, many studies suggested that ultimate axial strain can also be related to the lateral confining pressure (e.g. [3,6,15,28,33,36,37,39]). In literature, some methods for predicting the ultimate strain of FRP-confined concrete cylinders have been proposed. Existing models can be classified into three categories as follows:

(a) Steel-based confined models (e.g. [1, 40]), Saadatmanesh et al. (1994) [1] assumed that:

$$\frac{\varepsilon_{cc}}{\varepsilon_{co}} = 1 + 5\left(\frac{f'_{cc}}{f'_{co}} - 1\right) \tag{9}$$

where ε_{co} is the strain in peak stress of unconfined concrete and ε_{cc} is axial strain at peak stress of the FRP-confined concrete.

(b) Empirical or analytical models (e.g. [10,21,24,29,30,36,39,41]), Teng et al. (2002) [21] proposed:

- For CFRP wrapped concrete:

$$\frac{\varepsilon_{cc}}{\varepsilon_{co}} = 2 + 15\left(\frac{f_l}{f'_{co}}\right) \tag{10}$$

- For design use:

$$\frac{\varepsilon_{cc}}{\varepsilon_{co}} = 1.75 + 10\left(\frac{f_l}{f'_{co}}\right) \tag{11}$$

(c) Recently, some models for predicting the axial stress and strain of FRP-confined concrete were suggested based on numerical method or plasticity analysis (e.g. [42,46]), whereas these models are often not suitable for direct use in design.

Proposed Equation

Figure 14 shows the relation between the strain enhancement ratio and the actual confinement ratio of the present test data. A linear relationship clearly exists. This diagram indicates that the axial strain of FRP-confined concrete can be related linearly to the actual confinement ratio. Based on regression of test data reported in Table 5, the axial strain of CFRP-wrapped concrete can be approximated by the following expression:

$$\frac{\varepsilon_{cc}}{\varepsilon_{co}} = 2 + 7.6\left(\frac{f_{l,eff}}{f'_{co}}\right) \tag{12}$$

Replacing $f_{l,eff}$ by f_l into Equation (12) the axial strain of FRP-confined concrete takes the form:

$$\frac{\varepsilon_{cc}}{\varepsilon_{co}} = 2 + 5.55\left(\frac{f_l}{f'_{co}}\right) \tag{13}$$

Given that ε_{cc} for concrete sufficiently confined by FRP is the ultimate strain ε_{cu}.

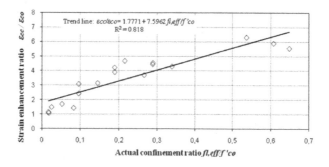

Figure 14. Strain enhancement ratio vs. actual confinement ratio

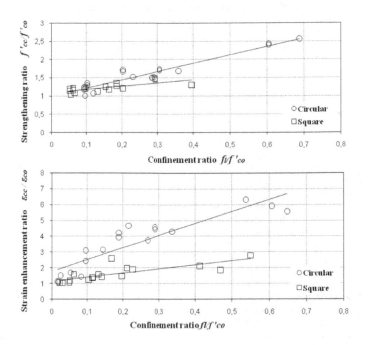

Figure 15. Strengthening ratio vs. confinement ratio and strain enhancement ratio vs. confinement ratio for the test results of this work

6.1.3. Validation of the Proposed Model

Using above model, the compressive strength and axial strain of FRP-confined specimens collected from other studies [6,36,47,48] were predicted as shown in Tables 5 and 6 which clearly exhibits excellent agreement between the experimental and predicted results. The present model is more accurate in predcting the compressive strength but less accurate in predicting the axial strain.

In Figure 15 the strengthening ratio-confinement ratio and the strain enhancement ratio-confinement ratio plots for the test results of this work (circular and square specimens) are shown, together with their respective linear regressions. From these Figures, it can be seen that the the axial confined compressive strength and the corresponding axial strain, approximately, increase linearly with the increase in confining lateral pressure for all types of section geometry. There is also a great distinction between the tendency of the results obtained for circular columns and those for square ones.

Specimen code	FRP Type	f'_{co} (Mpa)	E_{frp} (Gpa)	ε_{fu} (‰)	t_{frp} (mm)	d (mm)	f_l (Mpa)	k_1	$f'_{cc,theo,}$ (Mpa)	$f'_{cc,exp,}$ (Mpa)	$f'_{cc,theo,}/f'_{cc,exp,}$
Matthys et al. (2005) [6]											
k2	CFRP	32	198	11.9	0.585	400	6.891	1.6	43.027	54.30	0.792
k8	HFRP	32	120	9.6	0.492	400	2.833	1.6	36.534	44.40	0.822
Ilki et al. (2003) [47]											
CYL-5-1	CFRP	6.2	230	15	0.825	150	37.950	1.6	66.920	87.70	0.763
CYL-5-2	CFRP	6.2	230	15	0.825	150	37.950	1.6	66.920	82.70	0.809
Lam et al. (2006) [48]											
CI-M1	CFRP	41.1	250	15.2	0.165	152	8.250	1.6	54.300	52.60	1.032
CI-M3	CFRP	41.1	250	15.2	0.165	152	8.250	1.6	54.300	55.40	0.980
CII-M3	CFRP	38.9	247	15.2	0.33	152	16.302	1.6	64.983	65.80	0.987
Jiang et Teng (2007) [36]											
36	CFRP	38	240.7	15	1.02	152	48.456	1.6	115.530	129	0.895
39	CFRP	38	240.7	15	1.36	152	64.608	1.6	141.374	158.5	0.891
40	CFRP	37.7	260	15	0.11	152	5.644	1.6	46.731	48.50	0.963
41	CFRP	37.7	260	15	0.11	152	5.644	1.6	46.731	50.30	0.929
42	CFRP	44.2	260	15	0.11	152	5.644	1.6	53.231	48.10	1.106
43	CFRP	44.2	260	15	0.11	152	5.644	1.6	53.231	51.10	1.041
45	CFRP	44.2	260	15	0.22	152	11.289	1.6	62.263	62.90	0.989
46	CFRP	47.6	250.5	15	0.33	152	16.315	1.6	73.704	82.70	0.891
									Average: 0.926		
									Standard deviation: 0.101		
									Coefficient of variation (%): 10.90		

Table 5. Comparison of experimental and predicted results: compressive strength

Circular and Square Concrete Columns Externally Confined by CFRP Composite: Experimental
Investigation and Effective Strength Models

213

Specimen code	FRP type	ε_{co}	$\varepsilon_{cc,exp}$	k_2	$\varepsilon_{cc,theo}$	$\varepsilon_{cc,theo} / \varepsilon_{cc,exp}$
Matthys et al. (2005) [6]						
k2	CFRP	0.00280	0.0111	5.55	0.0089	0.806
k8	HFRP	0.00280	0.0059	5.55	0.0069	1.182
Ilki et al. (2003) [47]						
CYL-5-1	CFRP	0.00196	0.0910	5.55	0.0707	0.777
CYL-5-2	CFRP	0.00203	0.0940	5.55	0.0730	0.777
Lam et al. (2006) [48]						
CI-M1	CFRP	0.00256	0.0090	5.55	0.0079	0.885
CI-M3	CFRP	0.00256	0.0111	5.55	0.0079	0.718
CII-M3	CFRP	0.00256	0.0125	5.55	0.0110	0.885
Jiang et Teng (2007) [36]						
36	CFRP	0.00217	0.0279	5.55	0.0196	0.704
39	CFRP	0.00217	0.0354	5.55	0.0248	0.700
40	CFRP	0.00275	0.0089	5.55	0.0077	0.869
41	CFRP	0.00275	0.0091	5.55	0.0077	0.851
42	CFRP	0.00260	0.0069	5.55	0.0070	1.019
43	CFRP	0.00260	0.0088	5.55	0.0070	0.793
45	CFRP	0.00260	0.0102	5.55	0.0088	0.866
46	CFRP	0.00279	0.0130	5.55	0.0108	0.834
			Average:			0.845
			Standard deviation:			0.125
			Coefficient of variation (%):			14.80

Table 6. Comparison of experimental and predicted results: axial strain

6.2. Square Columns

6.2.1. Compressive Strength

The effective Lateral Confining Pressure

The effective lateral confining pressure f'_l can be defined as a function of the shape through the use of a confinement effectiveness coefficient k_e as:

$$f'_l = k_e \, f_l \tag{14}$$

were f_i is the lateral confining pressure provided by an FRP jacket and can be evaluated using Equation (1), with the columns diameter d replaced by the diagonal length of the square section. f_i now becomes an equivalent confining pressure provided by the FRP jacket to an equivalent circular columns. On the other hand, the effective FRP strain coefficient η' is defined as the ratio of the FRP tensile hoop strain at rupture in the square column tests ($\varepsilon_{h,rup}$) to the ultimate tensile strain from FRP tensile coupon tests (ε_{fu}):

$$\eta' = \frac{\varepsilon_{h,rup}}{\varepsilon_{fu}} \tag{15}$$

The effective FRP strain coefficient represents the degree of participation of the FRP jacket, and the friction between concrete and FRP laminate. Type bond, geometry, FRP jacket thickness, and type of resin affect the effective FRP strain coefficient. From the experimental results (Table 7), η' was 68 % on average for square bonded jackets.

Specimen code	f'_{co} (Mpa)	t_{cfrp} (mm)	E_{cfrp} (Gpa)	ε_{fu} (‰)	$\varepsilon_{h,rup}$ (‰)	d (mm)	f_l / f'_{co}	$f_{l,eff} / f'_{co}$	f'_{cc} / f'_{co}	ε_{co} (‰)	$\varepsilon_{cc} / \varepsilon_{co}$
SRCI.1L.$_1$	33.59	1	34	14	10.28	197.989	0.097	0.105	1.2051	4.29	1.249
SRCI.1L.$_2$	33.59	1	34	14	12.88	197.989	0.097	0.131	1.2373	4.29	1.585
SRCI.3L.$_1$	33.59	3	34	14	13.47	197.989	0.292	0.413	1.4534	4.29	2.093
SRCI.3L.$_2$	33.59	3	34	14	15.30	197.989	0.292	0.469	1.4713	4.29	1.825
SPCI.1L.$_1$	24.77	1	34	14	12.23	197.989	0.132	0.169	1.1167	2.17	2.571
SPCI.3L.$_1$	24.77	3	34	14	13.23	197.989	0.396	0.550	1.2931	2.17	2.788
SRCII.1L.$_1$	52.82	1	34	14	7.60	197.989	0.061	0.049	1.2009	4.07	1.066
SRCII.1L.$_2$	52.82	1	34	14	9.53	197.989	0.061	0.061	1.1484	4.07	1.594
SRCII.3L.$_1$	52.82	3	34	14	11.56	197.989	0.185	0.225	1.2755	4.07	1.909
SRCII.3L.$_2$	52.82	3	34	14	10.11	197.989	0.185	0.197	1.3406	4.07	1.476
SPCII.1L.$_1$	48.53	1	34	14	7.34	197.989	0.067	0.051	1.0822	3.38	1.192
SPCII.3L.$_1$	48.53	3	34	14	9.88	197.989	0.202	0.209	1.2003	3.38	1.988
SRCIII.1L.$_1$	63.79	1	34	14	5.78	197.989	0.051	0.031	1.1422	3.75	1.026
SRCIII.1L.$_2$	63.79	1	34	14	5.71	197.989	0.051	0.030	1.2043	3.75	1.037
SRCIII.3L.$_1$	63.79	3	34	14	7.16	197.989	0.153	0.115	1.2475	3.75	1.338
SRCIII.3L.$_2$	63.79	3	34	14	8.76	197.989	0.153	0.141	1.2478	3.75	1.402
SPCIII.1L.$_1$	59.53	1	34	14	3.97	197.989	0.054	0.022	1.0297	3.56	1.036
SPCIII.3L.$_1$	59.53	3	34	14	6.09	197.989	0.164	0.115	1.1818	3.56	1.387

Table 7. Data and results of CFRP confined square concrete specimens

Based on these observations, the effective equivalent lateral confining pressure f_l for square section, is given by:

-For square section:

$$f_l = \frac{2t_{frp}E_{frp}\varepsilon_{h,rup}}{\sqrt{2}b} = \frac{2t_{frp}E_{frp}\eta'\varepsilon_{fu}}{\sqrt{2}b} \qquad (16)$$

-For square section with round corners:

$$f_l = \frac{2t_{frp}E_{frp}\varepsilon_{h,rup}}{\sqrt{2}b - 2Rc(\sqrt{2}-1)} = \frac{2t_{frp}E_{frp}\eta'\varepsilon_{fu}}{\sqrt{2}b - 2Rc(\sqrt{2}-1)} \qquad (17)$$

Confinement Effectiveness Coefficient "k_e"

For the determination of the effectiveness factor k_e it can be assumed that, in the case of a circular cross-section, the entire concrete core is effectively confined, while, for the square section there is a reduction in the effectively confined core that can be assumed, analogously with the case of concrete core confined by transverse steel stirrups [17], in the form of a second-degree parabola with an initial tangent slope of 45°. For a square section wrapped with FRP (Figure 16) and with corners rounded with a radius Rc, the parabolic arching action is again assumed for the concrete core where the confining pressure is fully developed. Unlike a circular section, for which the concrete core is fully confined, a large part of the cross-section remains unconfined. Based on this observation, it is possible to obtain the area of unconfined concrete A_u, as follows:

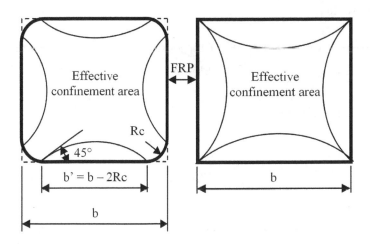

Figure 16. Effectivelly confined core for square sections

- For square section:

$$A_u = 4\left(\frac{b^2}{6}\right) = \frac{2b^2}{3}$$

(18)

- For square section with round corners:

$$A_u = 4\left(\frac{b'^2}{6}\right) = \frac{2b'^2}{3}$$

(19)

The confinement effectiveness coefficient k_e is given by the ratio of the effective confinement area A_e to the total area of concrete enclosed by the FRP jacket, A_c, as follows:

$$k_e = \frac{A_e}{A_c} = \frac{(A_c - A_u)}{A_c} = 1 - \frac{A_u}{(A_g - A_s)} = 1 - \frac{A_u}{A_g(1 - \rho_{sc})}$$

(20)

Where A_g is the gross area of column section, and ρ_{sc} is the cross-sectional area ratio of longitudinal steel.

By substituting the expression (18) or (19) into (20), the confinement effectiveness coefficient k_e is therefore given by:

- For square section:

$$k_e = 1 - \frac{2b^2}{3A_g(1 - \rho_{sc})}$$

(21)

- For square section with round corners:

$$k_e = 1 - \frac{2b'^2}{3A_g(1 - \rho_{sc})}$$

(22)

Proposed Equation

Base on the linear equation previously proposed by Richart et al. (1929) [22] for uniformly confined concrete, the proposed model employs similar approach with several modifications accounting for the effect of the shape, effective FRP strain and effective confinement.

The compressive strength of a square FRP-confined concrete column is proposed to be a simple modification of Equation (7) by the introduction of a confinement effectiveness coefficient denoted k_e. Thus,

$$\frac{f'_{cc}}{f'_{co}} = 1 + k_1 k_e \frac{f_l}{f'_{co}} \tag{23}$$

Where $k_e f_l / f'_{co}$ is the effective confinement ratio. The coefficient k_1 was taken as 1.60, which was suggested for uniformly confined concrete. Considering the known values of the product of the parameters k_1 and k_e as found from expression (23) for the tested columns of this work, the values of k_e were deduced, and were on average equal to 0.36. Finally, the equation proposed for the confined concrete strength is:

$$f'_{cc} = f'_{co} + 0.58 f_l \tag{24}$$

6.2.2. Axial Strain at Peak Stress

Similarly to the compressive strength, the axial strain at peak stress is proposed to be given by the following equation in which a different confinement effectiveness coefficient, k_{e2}, is introduced:

$$\frac{\varepsilon_{cc}}{\varepsilon_{co}} = 2 + k_2 k_{e2} \left(\frac{f_l}{f'_{co}} \right) \tag{25}$$

In Equation (25), f_l is the confining pressure in an equivalent circular column given by Equation (16) for square section, while $k_2 = 5.55$ and $k_{e2} = 0,72$. The equation proposed for the axial strain is:

$$\varepsilon_{cc} = \varepsilon_{co} \left[2 + 4 \left(\frac{f_l}{f'_{co}} \right) \right] \tag{26}$$

6.2.3. Comparison Between Proposed Model and Existing Test Data

Tables 8 and 9 show comparisons between the predictions of the proposed model and the experimental results collected from other studies [49,50,51,52] for the compressive strength and the axial strain at peak stress of FRP-confined concrete in square sections. Clearly, the present model is more accurate in predicting the compressive strength but less accurate in predicting the axial strain. Accurate predictions of the axial strain are an issue that will require a great deal of further research.

Specimen code	FRP type	f'_{co} (Mpa)	t_{frp} (mm)	E_{frp} (Gpa)	ε_{fu} (‰)	b (mm)	Rc (mm)	d (mm)	f_l (Mpa)	f'_{cc} (Mpa)	$f'_{cc.théo}$	$f'_{cc.théo}/f'_{cc.exp}$
Demers and Neale (1994) [49]												
-	CFRP	32.3	0.9	25	15.2	152	5	210.818	2.206	34.1	33.579	0.984
-	CFRP	42.2	0.9	25	15.2	152	5	210.818	2.206	45.99	43.479	0.945
-	CFRP	42.2	0.9	25	15.2	152	5	210.818	2.206	45.7	43.479	0.951
Lam and Teng (2003[b]) [51]												
S1R$_{15}$	CFRP	33.7	0.165	257	17.58	150	15	199.705	5.076	35	36.644	1.046
S2R$_{15}$	CFRP	33.7	0.33	257	17.58	150	15	199.705	10.15	50.4	39.589	0.785
Rochette (1996) [50]												
2B	CFRP	42	0.9	82.7	15	152	5	210.818	7.202	39.4	46.177	1.172
2D1	CFRP	42	0.9	82.7	15	152	25	194.249	7.816	42.1	46.533	1.105
2D2	CFRP	42	0.9	82.7	15	152	25	194.249	7.816	44.1	46.533	1.055
2G1	CFRP	42	0.9	82.7	15	152	38	183.480	8.275	47.3	46.799	0.989
2G2	CFRP	42	0.9	82.7	15	152	38	183.480	8.275	50.4	46.799	0.928
2C	CFRP	43.9	1.5	82.7	15	152	5	210.818	12.003	44.1	50.862	1.153
2E	CFRP	43.9	1.2	82.7	15	152	25	194.249	10.422	50.8	49.944	0.983
6A	AFRP	43	1.26	13.6	16.9	152	5	210.818	1.868	50.8	44.083	0.867
6D	AFRP	43	5.04	13.6	16.9	152	5	210.818	7.472	54.3	47.334	0.871
6E	AFRP	43	1.26	13.6	16.9	152	25	194.249	2.027	51.2	44.175	0.862
6F	AFRP	43	2.52	13.6	16.9	152	25	194.249	4.055	51.2	45.351	0.885
6G	AFRP	43	3.78	13.6	16.9	152	25	194.249	6.082	53.2	46.527	0.874
6H	AFRP	43	5.04	13.6	16.9	152	25	194.249	8.110	55.2	47.703	0.864
6I	AFRP	43	2.52	13.6	16.9	152	38	183.480	4.293	50.9	45.490	0.893
6J	AFRP	43	3.78	13.6	16.9	152	38	183.480	6.439	52.7	46.735	0.886
Benzaid (2010) [52]												
P300-R0-1P$_1$	GFRP	54.8	1.04	23.8	21.2	100	0	141.421	5.046	54.50	57.726	1.059
P300-R0-1P$_2$	GFRP	54.8	1.04	23.8	21.2	100	0	141.421	5.046	56.60	57.726	1.019
P300-R0-1P$_3$	GFRP	54.8	1.04	23.8	21.2	100	0	141.421	5.046	57.20	57.726	1.009
P300-R8-1P$_1$	GFRP	54.8	1.04	23.8	21.2	100	8	134.793	5.294	58.85	57.870	0.983
P300-R16-1P$_1$	GFRP	54.8	1.04	23.8	21.2	100	16	128.166	5.568	60.56	58.029	0.958

Average: 0.966

Standard deviation: 0.097

Coefficient of variation (%): 10.04

Table 8. Performance of proposed model: compressive strength

Circular and Square Concrete Columns Externally Confined by CFRP Composite: Experimental
Investigation and Effective Strength Models

219

Specimen code	FRP type	ε_{co}	$\varepsilon_{cc,exp}$	$k_2\,k_{e2}$	$\varepsilon_{cc,theo}$	$\varepsilon_{cc,theo}\,/\,\varepsilon_{cc,exp}$
Demers and Neale (1994) [49]						
1	CFRP	0.002	0.004	4	0.0045	1.136
2	CFRP	0.002	0.0035	4	0.0044	1.262
3	CFRP	0.002	0.0035	4	0.0044	1.262
Lam and Teng (2003[b]) [51]						
S1R$_{15}$	CFRP	0.001989	0.004495	4	0.0051	1.151
S2R$_{15}$	CFRP	0.002	0.0087	4	0.0064	0.736
Rochette (1996) [50]						
2B	CFRP	0.003	0.0069	4	0.0080	1.167
2D1	CFRP	0.003	0.0094	4	0.0082	0.875
2D2	CFRP	0.003	0.0089	4	0.0082	0.925
2G1	CFRP	0.003	0.0108	4	0.0083	0.774
2G2	CFRP	0.003	0.0116	4	0.0083	0.721
2C	CFRP	0.003	0.0102	4	0.0092	0.909
2E	CFRP	0.003	0.0135	4	0.0088	0.655
6A	AFRP	0.003	0.0106	4	0.0065	0.615
6D	AFRP	0.003	0.0124	4	0.0080	0.652
6E	AFRP	0.003	0.0079	4	0.0065	0.831
6F	AFRP	0.003	0.0097	4	0.0071	0.735
6G	AFRP	0.003	0.011	4	0.0076	0.699
6H	AFRP	0.003	0.0126	4	0.0082	0.655
6I	AFRP	0.003	0.0096	4	0.0071	0.749
6J	AFRP	0.003	0.0118	4	0.0077	0.660
Benzaid (2010) [52]						
P300-R0-1P$_1$	GFRP	0.0025	0.0088	4	0.0059	0.672
P300-R0-1P$_2$	GFRP	0.0025	0.0090	4	0.0059	0.657
P300-R0-1P$_3$	GFRP	0.0025	0.0098	4	0.0059	0.604
P300-R8-1P$_1$	GFRP	0.0025	0.0091	4	0.0059	0.655
P300-R16-1P$_1$	GFRP	0.0025	0.0098	4	0.0060	0.613
			Average:			0.815
			Standard deviation:			0.214
			Coefficient of variation (%):			26.30

Table 9. Performance of proposed model: axial strain

7. Conclusions

The results of this investigation have confirmed previous observations on the efficiency of confining FRP wraps. More specifically, the following concluding remarks can be made.

• It is evident that in all cases the presence of external CFRP jackets increased the mechanical properties of PC and RC specimens, in different amount according to the number of composite layers, the concrete properties and the cross-section shape.

• The failure of CFRP wrapped specimens occurred in a sudden and 'explosive' way preceded by typical creeping sounds. For cylindrical specimens, the fiber rupture starts mainly in their central zone, then propagates towards other sections. Regarding confined concrete prisms, failure initiated at or near a corner, because of the high stress concentration at these locations,

• CFRP strengthened specimens showed a typical bilinear trend with a transition zone. On overall, both ultimate compressive strength and ultimate strain are reached at the same point and are variably enhanced depending on the effect of other parameters.

• The efficiency of the CFRP confinement is higher for circular than for square sections, as expected. The increase of ultimate strength of sharp edged sections is low, although there is a certain gain of load capacity and of ductility.

• The CFRP confinement on low-strength concrete specimens produced higher results in terms of strength and strains than for high-strength concrete similar specimens. Therefore, the effect of CFRP confinement on the bearing and deformation capacities decreases with increasing concrete strength;

• Increasing the amount of CFRP sheets produce an increase in the compressive strength of the confined column but with a rate lower compared to that of the deformation capacity.

• In existing models for FRP-confined concrete, it is commonly assumed that the FRP ruptures when the hoop stress in the FRP jacket reaches its tensile strength from either flat coupon tests which is herein referred to as the FRP material tensile strength. However, experimental results show that the FRP material tensile strength was not reached at the rupture of FRP in FRP-confined concrete and specimen's failure occurs before the FRP reached their ultimate strain capacities. The failure occurs prematurely and the circumferential failure strain was lower than the ultimate strain obtained from standard tensile testing of the FRP composite. This phenomenon considerably affects the accuracy of the existing models for FRP-confined concrete. So on the basis of the effective lateral confining pressure of composite jacket and the effective circumferential FRP failure strain a new equations were proposed to predict the strength of FRP-confined concrete and corresponding strain for each of the cross section geometry used, circular and square. Further work is required to verify the applicability of the proposed models over a wider range of geometric and material parameters, to improve theirs accuracy (particularly that of the axial strain at peak stress) and to place theirs on a clear mechanical basis. Both additional tests and theoretical investigation are needed.

Acknowledgements

Authors thankfully acknowledge the support of Sika France S.A (Saint-Grégoire, Rennes) for providing the fiber-reinforced polymer materials.

Author details

Riad Benzaid[1*] and Habib-Abdelhak Mesbah[2]

*Address all correspondence to: benzaid_riad@yahoo.fr

1 L.G.G., Jijel University- B.P. 98, Cité Ouled Issa, Algeria

2 L.G.C.G.M., INSA of Rennes, France

References

[1] Saadatmanesh, H., Ehsani, M. R., & Li, M. W. (1994). Strength and ductility of concrete columns externally reinforced with composites straps. *ACI Structural journal,* 91(4), 434-447.

[2] Mirmiran, A., Shahawy, M., Samaan, M., & El Echary, H. (1998). Effect of column parameters on FRP-confined concrete. *ASCE Journal of Composites for Construction,* 2(4), 175-185.

[3] Shehata, I. A. E. M., Carneiro, L. A. V., & Shehata, L. C. D. (2002). Strength of short concrete columns confined with CFRP sheets. *RILEM Materials and Structures,* 35, 50-58.

[4] Chaallal, O., Hassen, M., & Shahawy, M. (2003). Confinement model for axially loaded short rectangular columns strengthened with FRP polymer wrapping. *ACI Structural Journal,* 100(2), 215-221.

[5] Campione, G., Miraglia, N., & Papia, M. (2004). Strength and strain enhancements of concrete columns confined with FRP sheets. *Journal of Structural Engineering and Mechanics,* 18(6), 769-790.

[6] Matthys, S., Toutanji, H., Audenaert, K., & Taerwe, L. (2005). Axial load behavior of large-scale columns confined with fiber-reinforced polymer composites. *ACI Structural Journal,* 102(2), 258-267.

[7] Wu, G., Lu, Z. T., & Wu, Z. S. (2006). Strength and ductility of concrete cylinders confined with FRP composites. *Construction and Building Materials,* 20, 134-148.

[8]   Almusallam, T. H. (2007). Behavior of normal and high-strength concrete cylinders confined with E-glass/epoxy composite laminates. *Composites part B*, 38, 629-639.

[9]   Benzaid, R., Chikh, N. E., & Mesbah, H. (2008). Behaviour of square concrete columns confined with GFRP composite wrap. *Journal of Civil Engineering and Management*, 14(2), 115-120.

[10] Rousakis, T. C., & Karabinis, A. I. (2008). Substandard reinforced concrete members subjected to compression: FRP Confining Effects. *RILEM Materials and Structures*, 41(9), 1595-1611.

[11] Benzaid, R., Chikh, N. E., & Mesbah, H. (2009). Study of the compressive behavior of short concrete columns confined by fiber reinforced composite. *Arabian Journal for Science and Engineering*, 34(1B), 15-26.

[12] Benzaid, R., Mesbah, H., & Chikh, N. E. (2010). FRP-confined concrete cylinders: axial compression experiments and strength model. *Journal of Reinforced Plastics and Composites*, 29(16), 2469-2488.

[13] Piekarczyk, J., Piekarczyk, W., & Blazewicz, S. (2011). Compression strength of concrete cylinders reinforced with carbon fiber laminate. *Construction and Building Materials*, 25, 2365-2369.

[14] De Lorenzis, L., & Tepfers, R. (2001). A comparative study of models on confinement of concrete cylinders with FRP composites. Division of Building Technology, work No. 46, Publication 01:04. Chalmers University of Technology, Sweden. 81p.

[15] De Lorenzis, L., & Tepfers, R. (2003). A comparative study of models on confinement of concrete cylinders with fiber-reinforced polymer composites. *ASCE Journal of Composites for Construction*, 7(3), 219-237.

[16] Park, R., & Paulay, T. (1975). Reinforced concrete structures. John Wiley and Sons, N.Y., U.S.A. 800 p.

[17] Mander, J. B., Priestley, M. J. N., & Park, R. (1988). Theoretical stress-strain model for confined concrete. *ASCE Journal of Structural Engineering*, 114(8), 1804-1826.

[18] Cusson, D., & Paultre, P. (1995). Stress-strain model for confined high-strength concrete. *ASCE Journal of Structural Engineering*, 121(3), 468-477.

[19] Youssef, M.N., Feng, M.Q., & Mosallam, A.S. (2007). Stress-strain model for concrete confined by FRP composites. *Composites: Part B*, 38, 614-628.

[20] Rochette, P., & Labossière, P. (2000). Axial testing of rectangular column models confined with composites. *ASCE Journal of Composites for Construction*, 4(3), 129-136.

[21] Teng, J. G., Chen, J. F., Smith, S. T., & Lam, L. (2002). FRP strengthened RC structures. John Wiley and Sons Ltd., Chichester, UK. 245p.

[22] Richart, F. E., Brandtzaeg, A., & Brown, R. L. (1929). The failure of plain and spirally reinforced concrete in compression. Bulletin No. 190, Engineering Experiment Station, University of Illinois, Urbana, USA.

[23] Mirmiran, A., & Shahawy, M. (1997). Behavior of concrete columns confined by fiber composites. *ASCE Journal of Structural Engineering,* 123(5), 583-590.

[24] Samaan, M., Mirmiran, A., & Shahawy, M. (1998). Model of confined concrete by fiber composites. *ASCE Journal of Structural Engineering,* 124(9), 1025-1031.

[25] Saafi, M., Toutanji, H. A., & Li, Z. (1999). Behavior of concrete columns confined with fiber reinforced polymer tubes. *ACI Materials Journal,* 96(4), 500-509.

[26] Spoelstra, M. R., & Monti, G. (1999). FRP-confined concrete model. *ASCE Journal of Composites for Construction,* 3(3), 143-150.

[27] Xiao, Y., & Wu, H. (2003). Compressive behavior of concrete confined by various types of FRP composite jackets. *Journal of Reinforced Plastics and Composites,* 22(13), 1187-1201.

[28] Karbhari, V. M., & Gao, Y. (1997). Composite jacketed concrete under uniaxial compression- verification of simple design equations. *ASCE Journal of Materials in Civil Engineering,* 9(4), 185-93.

[29] Miyauchi, K., Inoue, S., Kuroda, T., & Kobayashi, A. (1999). Strengthening effects of concrete columns with carbon fiber sheet. *Transactions of the Japan Concrete Institute,* 21, 143-150.

[30] Toutanji, H. (1999). Stress-strain characteristics of concrete columns externally confined with advanced fiber composite sheets. *ACI Materials Journal,* 96(3), 397-404.

[31] Thériault, M., & Neale, K. W. (2000). Design equations for axially-loaded reinforced concrete columns strengthened with FRP wraps. *Canadian Journal of Civil Engineering,* 27(5), 1011-1020.

[32] Lam, L., & Teng, J. G. (2002). Strength models for fiber-reinforced plastic confined concrete. *ASCE Journal of Structural Engineering,* 128(5), 612-623.

[33] Lam, L., & Teng, J. G. (2003a). Design-oriented stress-strain model for FRP confined concrete. *Construction and Building Materials,* 17, 471-489.

[34] Berthet, J. F., Ferrier, E., & Hamelin, P. (2006). Compressive behavior of concrete externally confined by composite jackets- part B: modeling. *Construction and Building Materials,* 20, 338-347.

[35] Teng, J. G., Huang, Y. L., Lam, L., & Ye, L. P. (2007). Theoretical model for fiber reinforced polymer-confined concrete. *ASCE Journal of Composites for Construction,* 11(2), 201-210.

[36] Jiang, T., & Teng, J. G. (2007). Analysis-oriented stress-strain models for FRP-confined concrete. *Engineering Structures,* 29, 2968-2986.

[37] Ilki, A. (2006). FRP strengthening of RC columns (Shear, Confinement and Lap Spli-
      ces). *In: Retrofitting of Concrete Structures by Externally Bonded FRPs, with Emphasis on
      Seismic Applications*, Lausanne, Swiss. Fib Bulletin 35, 123-142.

[38] Yang, X., Nanni, A., & Chen, G. (2001). Effect of corner radius on the performance of
      externally bonded reinforcement. *In: Proceedings of The Fifth International Symposium
      on Fiber Reinforced Polymer for Reinforced Concrete Structures (FRPRCS-5)*, Cambridge,
      London, 197-204.

[39] Vintzileou, E., & Panagiotidou, E. (2008). An empirical model for predicting the me-
      chanical properties of FRP-confined concrete. *Construction and Building Materials*, 22,
      841-854.

[40] Fardis, M. N., & Khalili, H. H. (1982). FRP-encased concrete as a structural material.
      Magazine of Concrete Research, 34(121), 191-202.

[41] Siddhawartha, M., Hoskin, A., & Fam, A. (2005). Influence of concrete strength on
      confinement effectiveness of fiber-reinforced polymer circular jackets. *ACI Structural
      Journal*, 102(3), 383-392.

[42] Shahawy, M., Mirmiran, A., & Beitelman, T. (2000). Tests and modeling of carbon-
      wrapped concrete columns. *Composites Part B*, 31(6), 471-480.

[43] Karabinis, A. I., & Rousakis, T. C. (2001). A model for the mechanical behaviour of
      the FRP confined columns. *In: Proceedings of The International Conference on FRP Com-
      posites in Civil Engineering*, Hong Kong, China, 317-326.

[44] Moran, D. A., & Pantelides, C. P. (2002). Variable strain ductility ratio for fiber rein-
      forced polymer-confined concrete. *ASCE Journal of Composites for Construction*, 6(4),
      224-232.

[45] Becque, J., Patnaik, A., & Rizkalla, S. H. (2003). Analytical models for concrete con-
      fined with FRP tubes. *ASCE Journal of Composites for Construction*, 7(1), 31-8.

[46] Malvar, L. J., Morrill, K. B., & Crawford, J. E. (2004). Numerical modeling of concrete
      confined by fiber-reinforced composites. *ASCE Journal of Composites for Construction*,
      8(4), 315-322.

[47] Ilki, A., Kumbasar, N., & Koç, V. (2003). Low and medium strength concrete mem-
      bers confined by fiber reinforced polymer jackets. *ARI The Bulletin of the Istanbul Tech-
      nical University*, 53(1), 118-123.

[48] Lam, L., Teng, J. G., Cheung, C. H., & Xiao, Y. (2006). FRP-confined concrete under
      axial cyclic compression. *Cement and Concrete Composites*, 28, 979-958.

[49] Demer, M., & Neale, K. W. (1994). Strengthening of concrete columns with unidirec-
      tional composite sheets. In: Mufti, A.A., Bakht, B. and Jaeger, L.G. (eds), *Development
      in Short and Medium Span Bridge Engineering'94. Proceedings of the fourth International
      Conference on Short and Medium Span Bridges*, Canadian Society For Civil Engineering,
      Montreal, Canada, 895-905.

[50] Rochette, P. (1996). Confinement de Colonnes Courtes en Béton de Section Carrée ou Rectangulaire avec des Matériaux Composites. *Maîtrise Es-Sciences Appliquées*, Université de Sherbrooke, Canada, 115 p. (In French).

[51] Lam, L., & Teng, J. G. (2003b). Design-oriented stress-strain model for FRP-confined concrete in rectangular columns. *Journal of Reinforced Plastics and Composites*, 22(13), 1149-1186.

[52] Benzaid, R. (2010). Contribution à l'étude des matériaux composites dans le renforcement et la réparation des eléments structuraux linéaires en béton. *Thèse de Doctorat, INSA de Rennes*, France, 280 p. (In French).

Permissions

The contributors of this book come from diverse backgrounds, making this book a truly international effort. This book will bring forth new frontiers with its revolutionizing research information and detailed analysis of the nascent developments around the world.

We would like to thank Dr. Martin A. Masuelli, for lending his expertise to make the book truly unique. He has played a crucial role in the development of this book. Without his invaluable contribution this book wouldn't have been possible. He has made vital efforts to compile up to date information on the varied aspects of this subject to make this book a valuable addition to the collection of many professionals and students.

This book was conceptualized with the vision of imparting up-to-date information and advanced data in this field. To ensure the same, a matchless editorial board was set up. Every individual on the board went through rigorous rounds of assessment to prove their worth. After which they invested a large part of their time researching and compiling the most relevant data for our readers. Conferences and sessions were held from time to time between the editorial board and the contributing authors to present the data in the most comprehensible form. The editorial team has worked tirelessly to provide valuable and valid information to help people across the globe.

Every chapter published in this book has been scrutinized by our experts. Their significance has been extensively debated. The topics covered herein carry significant findings which will fuel the growth of the discipline. They may even be implemented as practical applications or may be referred to as a beginning point for another development. Chapters in this book were first published by InTech; hereby published with permission under the Creative Commons Attribution License or equivalent.

The editorial board has been involved in producing this book since its inception. They have spent rigorous hours researching and exploring the diverse topics which have resulted in the successful publishing of this book. They have passed on their knowledge of decades through this book. To expedite this challenging task, the publisher supported the team at every step. A small team of assistant editors was also appointed to further simplify the editing procedure and attain best results for the readers.

Our editorial team has been hand-picked from every corner of the world. Their multi-ethnicity adds dynamic inputs to the discussions which result in innovative outcomes. These outcomes are then further discussed with the researchers and contributors who give their valuable feedback and opinion regarding the same. The feedback is then collaborated with the researches and they are edited in a comprehensive manner to aid the understanding of the subject.

Apart from the editorial board, the designing team has also invested a significant amount of their time in understanding the subject and creating the most relevant covers. They scrutinized every image to scout for the most suitable representation of the subject and create an appropriate cover for the book.

The publishing team has been involved in this book since its early stages. They were actively engaged in every process, be it collecting the data, connecting with the contributors or procuring relevant information. The team has been an ardent support to the editorial, designing and production team. Their endless efforts to recruit the best for this project, has resulted in the accomplishment of this book. They are a veteran in the field of academics and their pool of knowledge is as vast as their experience in printing. Their expertise and guidance has proved useful at every step. Their uncompromising quality standards have made this book an exceptional effort. Their encouragement from time to time has been an inspiration for everyone.

The publisher and the editorial board hope that this book will prove to be a valuable piece of knowledge for researchers, students, practitioners and scholars across the globe.

List of Contributors

Martin Alberto Masuelli
Instituto de Física Aplicada, CONICET. Cátedra de Química Física II, Área de Química Física, Facultad de Química, Bioquímica y Farmacia. Universidad Nacional de San Luis. Chacabuco 917 (CP: 5700), San Luis, Argentina

Eustathios Petinakis
CSIRO, Materials Science and Engineering, Melbourne, Australia
Department of Materials Engineering, Monash University, Melbourne, Australia

Long Yu and Katherine Dean
CSIRO, Materials Science and Engineering, Melbourne, Australia

George Simon
Department of Materials Engineering, Monash University, Melbourne, Australia

Mônica Regina Garcez
Federal University of Pelotas, Brazil

Leila Cristina Meneghetti
University of São Paulo, Brazil

Luiz Carlos Pinto da Silva Filho
Federal University of Rio Grande do Sul, Brazil

George C. Manos and Kostas V. Katakalos
Laboratory of Experimental Strength of Materials and Structures, Department of Civil Engineering, Aristotle University of Thessaloniki, Greece

Theodoros C. Rousakis
Lecturer, Democritus University of Thrace (D.U.Th.), Civil Engineering Department, Engineering Structures Section, Reinforced Concrete Lab, Greece

Manal K. Zaki
Department of Civil and Construction Engineering, Higher Technological Institute, 6th October Branch, Guiza Egypt

Riad Benzaid
L.G.G., Jijel University- B.P. 98, Cité Ouled Issa, Algeria

Habib-Abdelhak Mesbah
L.G.C.G.M., INSA of Rennes, France